Oliver Graham-Jones

ZOO TAILS

BANTAM BOOKS

LONDON • TORONTO • SYDNEY • AUCKLAND • JOHANNESBURG

TRANSWORLD PUBLISHERS
61–63 Uxbridge Road, London W5 5SA
www.transworldbooks.co.uk

Transworld is part of the Penguin Random House group of companies
whose addresses can be found at global.penguinrandomhouse.com

Penguin
Random House
UK

First published in Great Britain in 2001 by Bantam Press
an imprint of Transworld Publishers
Bantam edition published 2002
Bantam edition reissued 2015

Copyright © Oliver Graham-Jones 2001
Map © Neil Gower 2001
Illustrations © Peter Bailey 2001

Oliver Graham-Jones has asserted his right under the Copyright,
Designs and Patents Act 1988 to be identified as the author of this work.

A CIP catalogue record for this book
is available from the British Library.

ISBN
9780857502605

Typeset in Granjon by Falcon Oast Graphic Art Ltd.
Printed in Great Britain by Clays Ltd, St Ives plc

Penguin Random House is committed to a sustainable
future for our business, our readers and our planet. This book is made
from Forest Stewardship Council® certified paper.

MIX
Paper from
responsible sources
FSC® C018179

1 3 5 7 9 10 8 6 4 2

CONTENTS

ACKNOWLEDGEMENTS

THROUGHOUT MY CAREER AT LONDON ZOO, THE zookeepers helped me a very great deal, and I am grateful for their friendship and skills. I was particularly impressed with the way many of them made the transition from their keeping duties to the new and highly technical world of the zoo hospital. In this respect, I would like to mention the hospital superintendent, Alec Wilson, and his second in command, Tony Fitzgerald, both now sadly deceased.

Other friends have also helped me, in particular my good friend Dr Patrick Moore, whose wisdom and persistence ensured publication of *Zoo Tails*. Patrick is a great and busy man, but he has always found time to encourage countless people, including me.

Among zoologists, Desmond Morris understood my problems very well, and I acknowledge his friendship with gratitude.

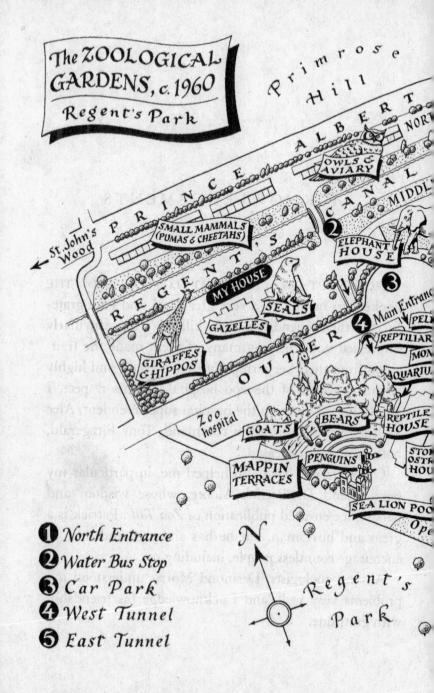

The ZOOLOGICAL GARDENS, c. 1960
Regent's Park

Primrose Hill

PRINCE ALBERT

NORTH

OWLS & AVIARY

CANAL

MIDDL...

← St. John's Wood

REGENT'S

SMALL MAMMALS (PUMAS & CHEETAHS)

❷ ELEPHANT HOUSE

❸

MY HOUSE

SEALS

GAZELLES

OUTER

❹ Main Entrance

PEL...

REPTILIAR...

MON...

GIRAFFES & HIPPOS

AQUARIU...

Zoo hospital

GOATS

BEARS

REPTILE HOUSE

STO...

OSTRI...

HOU...

MAPPIN TERRACES

PENGUINS

SEA LION POO...

Ope...

❶ North Entrance
❷ Water Bus Stop
❸ Car Park
❹ West Tunnel
❺ East Tunnel

N

Regent's Park

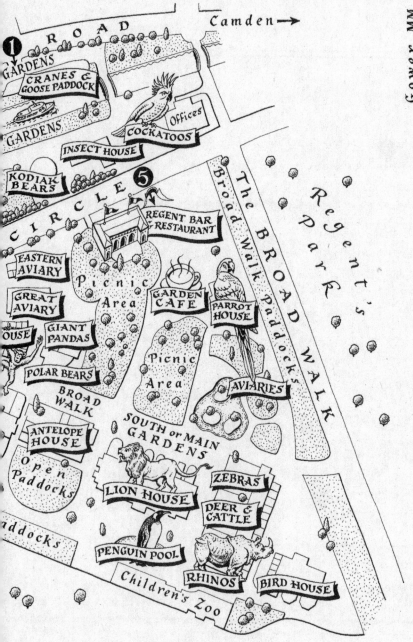

FOREWORD
by Desmond Morris

OLIVER GRAHAM-JONES IS A REMARKABLE MAN, AS YOU will soon discover as you read this collection of amazing anecdotes. In them, he describes the dramas and the emergencies, the triumphs and the tragedies, that he encountered during his days in charge of veterinary matters at London Zoo. Ordinary vets, working with domestic animals, encounter problems that are difficult enough, but Oliver was an extraordinary vet with an extraordinary collection of patients. And the almost daily problems he faced were equally extraordinary.

What would you do, for instance, if you were suddenly faced with an escaped bear, or an injured elephant, or an angry prime minister demanding that you place a live leopard on the cabinet table at Number 10 Downing Street? Oliver met all such challenges with a resourcefulness that had to be seen to be believed.

I first met Oliver in 1956, when my wife and I arrived

at London Zoo to start a new television unit, making films and TV programmes about the animals there. Although we were, in a sense, intruders into his world, he was kindness itself and helped us in every way possible. Our offices were in his old sanatorium building. He was busy moving into the splendid new animal hospital, which he had bullied the zoo into building for him. The animals owed him a great debt of gratitude for this, because he had at last managed to force the horribly old-fashioned zoo out of its Victorian past. Now he had a proper operating theatre, suitable quarantine enclosures and hygienic conditions for sick animals, all of which had been badly needed for a long time. Single-handedly, he dragged the zoo into the high-tech world of modern veterinary practice. It's astonishing to think that he was the first full-time vet the zoo had ever had.

Some of his best stories come from the earlier part of his career, when matters were still much more primitive. My favourite one is concerning a bizarre encounter with Winston Churchill. Oliver's rapidly developing skills at dealing with unexpected challenges proved especially useful when faced with the great man's eccentric whims. This and Oliver's other Zoo Tails will raise many a smile and, along the way, a greater appreciation of what it is to play doctor to the most exotic collection of patients in the world.

PROLOGUE

THE NIGHT SKY PALED, AND TROPICAL BIRDS BEGAN TO screech their greetings to the dawn. Soon the monkeys would join in with their worried chatter, an elephant would trumpet and the cacophony of the jungle would serve as a prelude to my working day.

Sleep-drugged, I switched off the alarm clock and marvelled, not for the first time, what on earth had possessed me to opt for a life where this morning chorus was as much part of the routine as, say, the clatter of milk bottles on a suburban doorstep.

As I tottered into the flat's bathroom and started to shave, I heard the roar of a lion – compulsively impressive. It was odd to reflect that Camden Town lay less than a mile away!

Over 1,000 mammals, birds, reptiles and fish of more than 250 different species, and I was their surgeon and medical adviser. It seemed a truly daunting prospect. How had it all begun, my association with the world's largest living museum, the world's biggest collection of wild animals in captivity? Yesterday's sick call had produced a panel of patients ranging from an ostrich to a spider monkey. The day before I had attended a python and the day before that a bear. Just what would today's tally be?

Adjusting my stiffly starched white coat – the symbol of office – I stepped into the morning air and walked towards the strangest, and sometimes noisiest, hospital in the world; a hospital where the surgeon is often in far greater danger than the patient; a hospital where you have to catch your tiger before bringing it to the table; the animal hospital in London Zoo.

INTRODUCTION

HIGH IN THE TREES AT THE BOTTOM OF OUR GARDEN, A thrush sang gloriously through the hush of the summer evening. A seven-year-old, lying pensively on his stomach, I turned to my mother in the deckchair beside me and asked, 'Why does that little bird sing?'

Mother lowered her knitting, and I could sense her sudden bafflement. 'Why does it sing?' she slowly repeated. 'Well, I don't really know.' Then, brightening, she added, 'Perhaps because it's so happy, dear.'

'But it can't sing *only* when it feels happy,' I answered, unimpressed. 'What does it sing about when it's miserable?'

My mother replied, 'You *do* ask such very odd questions.'

For the first time in my life I realized that parents were not an omniscient, infallible source of information, and the thought really startled me. I wasn't happy until I managed to ferret out the real scientific explanation for the thrush's exquisite vocal powers.

This was the start of my interest in species other than my own, an interest that stemmed from curiosity and owed little to sentiment. Today, people often say to me, 'You must always have loved animals,' and look a little shocked when I beg to disagree. Interest in a creature, and the desire to remedy its disabilities, should not be confused with sentimentality; for the patient's sake, the surgeon must always remain uninvolved.

I must have been about nine when my mood of enquiry into the behaviour of the living broadened to include reflections on the dead. There was nothing morbid about this, it was just that I wondered *why*.

Why should a field mouse, retrieved with difficulty from the cat's jaw, pass away within minutes of its rescue? It bore no signs of injury; it had only been heavily teased, so why should it die? My mother said something about a heart attack – 'The fright was too much for it' – but this set me on the path to further questioning. *Why* did the mouse's fright make its heart stop beating? *Why* couldn't its heart be started up again? A clock that had stopped could be rewound, I said, so why couldn't the mouse's heart? I really must have been a portentous bore!

* * *

At the time when all these exercises in natural history-cum-physiology were taking place, I lived in a small but very pleasant house on Birmingham's Harborne Estate with my parents, brother and sister. To us children the place was very special, and even now I can recall the great sense of outrage and personal loss that swept over us when we heard that we were to move and settle elsewhere.

The house had grown too small for our expanding family, but the fact that it was almost bursting at the seams with the stress of our explosive energies, or that there was nowhere for the grown-ups to get a moment's peace, didn't worry us. Extremely content in the home we had known since birth, the threat to remove us from our cocoon, with all the security it represented, filled us with terror.

However, when the dread day arrived, we found that things weren't so bad after all. Our new quarters, only half a mile from World's End Lane – nowadays engulfed by the concrete ocean of development – offered us even greater scope than 8, Park Edge for following rural activities. What with our encounters with the cattle, grazing in lush fields where today not a single blade of grass can be seen, our fishing for tiddlers in the nearby stream and our long walks with Father – a beloved man – through the meadows and woods, we enjoyed the life of country children and my interest in animals became more and more absorbing. In fact, it was at 15, Milford Road, Harborne, that I had my first 'veterinary' experience: treating my dog for injuries sustained in heroic battle.

Our dog was a mongrel Irish terrier, with a nervous, sensitive nose and soft brown eyes; he impressed all who saw him with his gentleness. Unfortunately perhaps for him, his looks belied his nature. Pat was a battler, a sort of canine d'Artagnan; a swaggering, carefree blade with courage and style. Impulsive in his temperament, the only fault in his fighting was that invariably he would pick a dog of double his size and weight as an opponent. This seemed almost a point of honour with him and he bore the inevitable consequences with equanimity, painful though they must have been.

So when we moved to Milford Road it was no surprise when he challenged the biggest dog in the village to assert his – far-reaching – claims of domain. Nor that, undeterred by honourable scars, he returned to the attack the following day ... and then again a couple of days later. It was a process that became repetitive, and it was deplored by everyone in the family except me, Pat's titular master. *I* thought him stupendous.

Two or three times a week our dog would engage with his rival, contesting with him for mastery and always getting the worst of it. Each time he would return home extremely late, as if anxious not to disturb my parents and let the sun go down upon their well-merited wrath. On such occasions, as he seemed to expect, I would creep downstairs in my pyjamas to let him quietly into the kitchen, and bathe and cleanse his wounds as best I knew. Every time he would prevent the true worth of the treatment being accurately assessed, for long before his injuries had a chance to heal, he would be off for the next round, to fight and fight again!

Eventually, however, our mutual conspiracy of silence failed. Pat's fights became a scandal and caused public complaint. He wrecked the local paper shop one morning by choosing to fight his pet aversion inside while the commuters tried to catch their bus. After that my parents presented him to the postman, which neatly solved two problems, the first to rid the house of a canine embarrassment, and the second to help guard the postman from other dogs on his rounds. I had my first lesson in emotional disassociation – so vital to one destined to become a veterinary surgeon.

Curiosity about why a wild bird should sing; dramatization over the plight of a rabbit, whose sentiments I equated with those of man; sympathy and comradeship for my embattled dog – it would be wrong to say that any single one of these factors was decisive in my choice of career, nor was there any individual revelation that caused me to think, I am going to be a vet! The process contrived to grow on me, almost without my noticing it.

At grammar school, my interests in animals and their habits acquired me a certain fame or notoriety, which reached its peak when I made a home for white mice in my jacket pocket and Mother discovered their droppings. For the sake of my pets' personal safety I then made the hurried decision to take them with me to school, where I was able to provide them with a far roomier place to nest – inside the back of my desk.

For a while this arrangement worked famously, but the hour of reckoning came when a master asked me to produce an exercise book. I delved into the desk, fumbled in the back, and felt my face go oven-hot with embarrassment. All that remained of the book I unwillingly presented was a mass of soggy paper, pulped and perforated by a thousand bites. The master reached my side before I could drop the desk lid, and there, exposed to his startled gaze, were my original white mice and more, breeding like mad things, and with healthy appetites, too!

This incident marked the conclusion of my social study of the domesticated mouse. Feeling that the situation had got slightly out of hand, I agreed to get rid of them, and swapped them at the exchange rate of one full-grown mouse for two marbles, a baby mouse for one.

At the zoo, when faced with the problem of how to anaesthetize a gorilla, we produced a giant gas box, which we placed flush against the opening to the animal's den in the hope the gorilla would wander into it.

The gorilla is a cunning and highly suspicious customer, and doesn't lightly forsake the security of its familiar environment; much persuasion is needed before it will play the game according to your rules. So, by way of inducement, we placed a full-length mirror at the far end of the box. Immediately the animal saw itself reflected there, his wariness vanished and he responded in the way we wanted. As he entered the box, the slide fell shut behind him, sealing him off. He had been trapped by his own curiosity.

It was probably this self-same quality – curiosity – that made me decide, when still a fourth former, what career path I should follow in the years ahead. Curiosity, and a perverse desire to swim against the tide. For my decision was not likely to be enthusiastically received.

My brother was training as a doctor, my sister as a teacher, and Mama had a single-minded idea, inspired by I know not what, that I should be a dental surgeon. Indeed, she had entered my name for dental college long before, academically speaking, I had cut my own teeth, let alone nourished any ambitions of operating on anyone else's.

My schoolmasters also weren't particularly keen on my idea of following a veterinary career. Indeed, when I started to enquire how to set about it, I found that it was a mystery not only to me, but to adults as well.

This became particularly obvious when I was one day sent for by the headmaster. When, my conscience at rest, I told him that I wished to be a veterinary officer, his air of paternal assurance swiftly vanished. A vet? Why a vet? Wasn't I entering the dental profession? I explained my views and repeated my request. Could he tell me, please, what I would need to qualify? Did he know where I should write? Eventually the interview terminated with me none the wiser and the head quite foxed and a little irritated at being, for once, left without an answer.

I left the headmaster's presence with the lesson clear in my mind that if you wanted to find out anything worthwhile you had better do it yourself. But it was only after I had pestered the most unlikely sources for information that I finally tackled the only people who really knew about it: the Royal College of Veterinary Surgeons. And no sooner had I learned what I wanted to know, and obtained, after some discussion, an unenthused parental blessing, than I almost changed my mind and shelved the idea altogether. Paradoxically, it was my father who put me back on course . . .

My father, a journeyman silversmith by articles, who had suffered severely during the Depression, had started up in business as a shipfitter, using plastics, and was doing remarkably well. At the end of each ocean-crossing there were calls made on his services, from ships whose names now have an almost legendary fame, the *Queen Mary* and the *Queen Elizabeth* among them, and so brisk was the business that I was enlisted as a part-time worker during the school holidays.

Of course, all this activity, with its introduction to the big liners, seemed highly dramatic to one of my impressionable age. It was certainly lucrative when compared with my normal pocket-money allowance, and it required no undue concentration. My studies began to suffer and I toyed with the idea of becoming a shipfitter, too, until Father, sensing my feelings, put an end to it. Calling me into his office, he said, 'You come into this business only over my dead body!'

Father had seen how a man without qualifications, except his intelligence and capacity to work, could have his livelihood threatened, and then removed, by market pressures over which he exerted no control. He was determined that his son should not be similarly vulnerable, but should have what he had lacked: the equivalent of a university course in a specialized profession. So back I went to pursue my real ambitions.

Having overcome her initial incredulity at my failure to be enthused by dentistry, my mother was a tower of strength during the months that followed. She fixed an interview for me at the Royal Veterinary College, to decide whether or not I might enrol as a student, and then, as ready to fight for my career as she had hitherto been to oppose it, she journeyed to London with me to give me moral support.

I entered the Royal Veterinary College and there, at the focus-point of my ambitions, so to speak, I found myself prey to the most awful homesickness and the most shaming doubts about my ability.

A place with the gloss, and much of the tang, of new white paint; a place of austerity, coldly gleaming in wan sunshine and with windows whose glassy stare seemed to cover every inch of the trim and weedless drive. Such was my first impression of the institution to which, for better or worse, I had been assigned for the next five years of my young life: the Royal Veterinary College.

Founded in the late eighteenth century, the college was undergoing a comprehensive rebuilding programme when

I started there, and its décor was severely contemporary in the 'hygienic', if rather sterile, fashion of the Thirties. Even the canteen, a gleaming mass of stainless steel, was in a chaste sort of way a thing of beauty, but the most visually impressive of all the spectacles the college had to offer was its magnificent dissecting theatre, which seemed as large as an aircraft hangar.

From the trestles, a large number of animals, and parts of animals, hung from chains, to be lowered or heightened for examination by the students. Exhibits included a comprehensive range of specimens, and dominating the scene was a very dead horse, pickled and ready for dissection. I regarded this with the sickness of despair, eyes streaming from the acrid fumes of formalin.

The programme was extremely comprehensive, and we were awed by all the things we would have to know. In some cases it was even more advanced than that judged appropriate for human medicine, for one would have to diagnose cases without recourse to the information a human patient can supply. Animal Husbandry, Anatomy, Physiology, Pharmacology, Pathology, Medicine, Surgery – all these subjects, and more, had to be assimilated during the five years of one's labours, and there were major examinations at annual intervals, which culminated in the fateful finals, taken in my case while still twenty-one.

The lecturers were men of great talent, and several of them were characters as well. We discussed their eccentricities with juvenile levity, but, in actual fact, we regarded them with little short of awe.

One of the greatest of professors to bestride the teaching scene affected the morning suit style – frock coat and striped trousers – that had been favoured by his tutor's tutor before him. With a fresh rose in his lapel, inches below the rim of his severely glossed winged collar, his sole concession to modern trends was his frenetic chain-smoking. Wearing – it's the only word for it – a succession of cigarettes beneath his bushy wicked-squire-type moustache, and replacing each one only when it had been burned down to the butt and started to singe his lips, he would hold the class mesmerized, waiting for his imminent conflagration. The major points of his discourse were emphasized by a finger adorned by nicotine and a monstrous ring. Yet, appearances notwithstanding, he was a truly splendid teacher, of international repute, and had written what was then the only encyclopedia of veterinary medicine. He was also a highly competent practitioner and demonstrator of the use of the surgeon's knife if required.

In 1939, the outbreak of war brought to a halt the stately minuet of donnish politics, and scattered us into the

English countryside. For the Royal Veterinary College, evacuation entailed a removal to the delightful Berkshire village of Streatley, where we spent the next few months before embarking on our finals. To my astonishment, I qualified at the very first attempt.

Flushed with the arrogance of youth that I now find so perplexing when encountered among the young people of today, I immediately abandoned all ideas of becoming assistant to an established veterinary surgeon, and purchased myself the nucleus of a practice. And then, in what may now appear to have been a rather quaint example of second-thoughts, I volunteered for the Army, and the Army said No.

The Italian campaign was about to begin, and the Army needed mules for transport work in the mountains. Until then understandably preoccupied with armoured warfare, the Army now had to find men who could handle animals by advertising for them in the newspapers. When I applied I was told that there were obstacles in the way. I was a married man in a reserved occupation, and, more important still, I happened to own a practice. My job was to keep animals fit for the battle of the land.

Several weeks went by before the Army relented, and then only because of my persistent pressure. I found myself in Italy in next to no time at all.

From the snug expanding practice in the pleasantest part of Surrey to the Field Remount Depot on the circuitous road to Rome, from sitting up with a sick cow in the calm of an English meadow to supplying teams of mules – the transition was sharp and far from pleasant. And yet, like so many others of my age group, I would not have missed my share of the mutual misery. I hated the life, but the experience made it worthwhile.

I danced with the local mayor's daughter, and saw the fleas dance with us – in her hair. I impounded sheep to feed the officers' mess, and was forced to pay for my cheek by the irate army officer responsible for the town. I slept in war-blasted ruins, waking to dust the snow off my sleeping bag, and curse the 'Sunny South' as a criminal fraud. I was eventually caught up by an old enemy: brucellosis. This causes excessive aborting among cattle and severe illness, with prolonged febrile attacks, in man. It sometimes takes fifteen years or more to run itself out, and it certainly took that long to run its course with me.

I had contracted the disease without knowing it, back in England as a student, where I had handled infected cows. It first manifested itself in a high temperature and a series of excruciating headaches. It resulted in my being sent out of the line for diagnosis.

After a Cook's tour of American hospitals I was

eventually invalided home and discharged. I had served for two years.

My departure from the Army left me rudderless. I missed my friends and the comradeship. As a survivor, I knew that I was 'one of the lucky ones', but like other survivors I was almost ashamed of my luck. Man is never really content; it takes an animal to be that!

It was perhaps because of this – a sense of guilt maybe at the contrast between my fate and that of others – that I began to have serious doubts about my civilian career. The practice was booming, but I wasn't sure it was all I really wanted. I began to feel as though I wasn't helping animals, but was merely cashing in on them by enhancing their productivity for the farmer.

It was in 1947 that I found a panacea for my doubts by deciding to invest part of the proceeds of the practice in a line that, at that time, was almost completely untried: I decided to finance and design an animal hospital. One of the first of its kind outside the teaching schools, I wanted to offer a first-class veterinary service, provided with up-to-date facilities for diagnosis and surgery – facilities that, otherwise, owners could not afford. By the time I had completed the place it had cost me all my savings and more.

It took five more years for disillusionment to set in; not with the product, but with the way people treated it. It became abundantly clear to me that the practice of the more enlightened veterinary skills was ultimately dependent on the depths of the owner's pocket, or rather on his willingness to dig therein, which is not always the same thing. Disappointed, and somewhat bitter, I decided to close the hospital, dispose of the practice and quit veterinary work altogether and enter my name for one of London's famous medical schools. From now on, I would switch to human medicine and try to qualify as a doctor.

A former professor at the Royal Veterinary College who remembered me as a student was genuinely concerned at my leaving the profession. One day, taking me aside, he told me that, owing to the Veterinary Surgeons' Act of 1948, London Zoo had decided to look for a resident veterinary officer, and he suggested I put my name forward for the post. While appreciating his kindness I told him I wasn't interested. Refusing to take my no as final, he tried again, and repeated his efforts until finally, really only to humour him, I agreed to attend an interview. I was confident I wouldn't be offered the job as, until then, my veterinary experience had been confined to agricultural animals and pets. I knew as much about wild animals as most of my contemporaries, and that was very little.

On the day of the interview, I entered the room nonchalantly, confident that the procedure would be over in a matter of minutes. Two and a half hours passed before I emerged, humbled and cut down to size. The questions put to me had revealed that my ignorance was even more abysmal than I had imagined. They had also triggered a grudging curiosity on my part regarding the nature of the job, in which, until then, I had been uninterested.

How did one administer medicine to a tiger? How could a tiger be brought to the operating table? Come to think of it, how did one diagnose the illness of a tiger?

The more I dwelt on these problems, the more I began to want the job. I found myself admitting that it presented a challenge, and could have been interesting, had they offered it to me. But they hadn't, and were certain not to, and that was the end of it.

Six weeks later I received a letter of appointment from the zoo. They offered me £800 a year, plus the occupancy of a small flat in the Zoological Gardens. Astonished, I immediately decided to accept.

I and the tiger were to make each other's acquaintance, after all, but first I would have to learn to catch the beast!

FIRST DAY

AS WE GROW UP WE HAVE MANY 'FIRST DAYS' IN OUR LIFE.
There is our first day in a nursery school, our first day in
an infant school, our first day in a primary school, our
first day in a secondary school and, for those lucky
enough to make the grade, our first day at university. My
first day at the Zoological Society of London was
terrifying.

I discovered that the joint appointment of Curator of
Mammals and Veterinary Officer was really a fund-
saving exercise to get two men for the price of one, and
also to comply with the 1948 Veterinary Surgeons' Act,
which required none other than a qualified veterinary
surgeon to diagnose or treat animals.

On the same day that I was appointed Curator of
Mammals and Veterinary Officer to London Zoo, John
Yealland was appointed Curator of Birds. Despite the

resonance of the titles we were, both of us, very much the new boys on the first day of term when we arrived, punctually, on the stroke of nine, at the office of the zoo superintendent, at that time the inimitable George Cansdale.

George was a most excellent superintendent, and a kindly man; he met me and said, 'Here is your office. Make your own job; you are the first full-time resident veterinary surgeon we have employed, and I have no idea what you want to do. In the first instance I suggest that you go on the rounds with the overseers to introduce you to the keeper staff and the various animal houses.'

I left George's office and went back to the main entrance hall of the society's headquarters building. There, standing stiffly to attention and looking immensely smart, stood overseer Vinall. The overseers of the zoo were equivalent to RSMs in the army; they had charge of a section with all its houses, animals and keepering staff. Overseer Vinall had been a gunnery sergeant major, and like so many war-time soldiers he had carried his discipline and deportment into civilian life. His uniform was immaculate, the cap badge polished and the brim shining. His long blue mackintosh was spotless and his shoes gleamed on feet that were precisely 45 degrees apart. As I approached him he came smartly to a salute and stamped his leg down to attention.

'Sah,' said Vinall. 'It is a pleasure to meet you. Here is your sick list, sah.' At which point he saluted again and automatically stamped to the 'at ease' position. I imagine he was waiting for me to say something brilliant. I took the sick list from him with a quivering hand, and lowered my eyes from his steely gaze to look at the contents of the list. What I saw burned holes in my brain and they remain there to this day. The list read:

1. 1 puff adder
2. 1 antelope
3. 1 crocodile

This was something of a culture shock. Only a week before I was in the leafy lanes of Surrey pleasantly going about my business in a country practice, seeing horses, cattle, pigs, dogs and cats, and chatting to farmers and so on. The brutal differences caused my knees to sag slightly.

In a fatuous, high-pitched voice I said, 'Thank you, Mr Vinall. Shall we do a round?'

All Vinall said was, 'Sah,' at which point he about-turned smartly and marched off, with me following.

Our first port of call was the crocodile. As we got to the reptile house, the keepers were lined up, looking very smart to create a good impression, no doubt, but mainly to see the fun. The head keeper was introduced to me formally by Vinall, and everyone waited to see what would happen next. On the march across from the main offices to the reptile house my mind was turbulent, but I decided that honesty was the best policy, so I said to the head keeper, 'As you know, I started work here today and I have a lot to learn. How about you telling me all about this sick crocodile and what you think is the matter with it, and I'll pitch in with my veterinary knowledge and see whether, between the two of us, we can get it right?'

This went down very well. There was a murmur of approval at the recognition that I admitted being totally ignorant. It also gave the head keeper a wonderful opportunity to display his vast store of knowledge of the practical day-to-day running of the reptile house, the animals, the keeper staff, nutrition, diet and disease processes. Between us we decided that, because the public were then allowed to feed the animals, the crocodile had got an extremely nasty bellyache, probably from swallowing coins, and that this needed medication. I had a good idea what I would give it, but no idea how it was going to be administered. The head keeper came to my rescue: 'You give us the medicine, sir, and we will see that it goes down.' Our interview concluded, Vinall and I were invited by the head keeper to take tea in the keepers' mess. This was the equivalent of an olive branch, a handshake, or the signing of a peace treaty.

After this we visited the puff adder and the antelope, where the introductions and results were similar. After this I was full up with mugs of keeper-brewed tea, which bore a distinct resemblance to the frightful stuff I used to drink during the war, brewed on top of a pot-bellied stove in a huge bucket and kept simmering for endless hours. It tasted quite awful, but to be offered a mug of tea, sweetened with condensed milk, was a mark of acceptance and a compliment, and it would have been rude to refuse.

Vinall then explained to me that it would be necessary for me to visit what in those days was called the society's sanatorium, otherwise known as the sanny.

We walked through the double gates which shielded the yard surrounding the sanny. I was very depressed at what I saw. The building was the old stables and haylofts, and they dated back to the days when the only transport in the gardens was horse and cart. When motorized transport arrived, a new department was built at the far side of the zoo, and the old stables and loose boxes were turned into the sanny. Downstairs there were the old stable double doors, and inside were a variety of creatures, hopefully all recuperating. On the first floor were what used to be hay and general food stores. These had been turned into cage rooms, and offices for the man in charge. He was the transport officer, and was really responsible for organizing the transport of animals in and outside the zoo, for the import, quarantine and transport arrangements of animals arriving from abroad. He lived about half a mile from the zoo in an old stables, which had been converted into a quarantine station.

He was a good man, but he knew nothing about sick animals, although his father had been a vet. He had an air of importance and seniority and wore a blue serge suit. He seemed to feel this gave him precedence over the uniformed staff. He did his best to be useful and helpful, but couldn't help being patronizing and off-hand. He clearly resented my presence, as now I was going to be in charge not only of much of the gardens, but of the sanatorium, too, which he had regarded as his personal empire – and of him!

The old sanny was so awful that, after a few months, I reported to the zoo council that there was no way anyone could offer a proper medical service to the animals in the society's collection unless they had a new hospital in which to work. Eventually a new hospital was built, and it is there to this day – a fully equipped building which is still an example to many. The old sanny was regarded as a dumping ground by the head keepers. They didn't want sick animals on their premises at all, and far less did they want them to die there. So when something got sick and failed to respond to simple remedies, it was transferred to the sanny to die, which it frequently did. Also staff in the old sanny were considered to be the most lowly. A transfer to a position in the sanny was looked on as a sort of punishment and loss of face.

At last came the end of my first day as the new Curator of Mammals and Veterinary Officer to the Zoological Society of London. It had been sheer hell, partly because I realized very quickly how little I knew and how much the keeper staff would need to, and did, help me. The most depressing thing perhaps was to see the old sanatorium, which was then the only place where I could attempt to apply modern veterinary science. The only redeeming feature of my first day was the warmth and kindness of the keepers once they had overcome the embarrassment of the first introduction and realized that I was honest enough to admit ignorance.

Keeper staff are frequently under-rated and under-paid, and they are a lesson in all sorts of ways to the ivory-tower inhabitants of main offices who think they run a zoo.

A POLAR BEAR SAGA

I WONDER HOW MANY PEOPLE REMEMBER THE FAMOUS smogs of yesteryear? For those of you who don't, let me explain. They were like thick acrid blankets of coloured fog which hung heavy in the air and reeked of various pollutants. I experienced some of the worst smogs of my life when I was a lad in the Midlands in the Thirties. They were yellowish in colour and stank heavily of the effluent from scores of smelting furnace chimneys in the Black Country, that area of the West Midlands dedicated solely to heavy industry. The presence of metal and acid in the fog was easily detectable and it had a choking effect that made you gasp and splutter. People would walk about with damp scarves around their faces to act as gas masks. Thousands of people were affected by lung disease and fatal pneumonia during those times. Tragically, 3,000 to 4,000 people died in the Midlands as a result of a particularly heavy smog.

Smogs in London were not dissimilar, but they didn't have the intense density of those in the Midlands. Nonetheless, London smogs were very heavy and smelled mostly of domestic chimneys, diesel and petrol fumes. They were also very thick, and in what you might call a five-star smog, visibility was down to a few feet. You could barely see the hand on your outstretched arm.

Imagine the effect of such a smog in any zoo, and in particular Regent's Park Zoo. Servicing animals in such a smog is extremely difficult. The keepers have difficulty getting to work from their homes, although, luckily, many of them lived within walking distance and their loyalty to their animals was always an inspiration to me. Keepers also find it difficult to get round the zoo in a smog and they cannot distribute the animals' food – which is done by car – as it is too difficult to drive until the smog lightens. One late October there was just such a smog.

During the winter the public are not admitted until 10 a.m., mainly because it is too dark before then. The keeper staff are normally on duty at 8 a.m., but in dreadful conditions like this they would be late.

I was in my flat at the zoo, which was located between the seals and the hippopotamus, getting ready for the day when the internal telephone pealed in a way that somehow made me aware there was trouble. An excited keeper shouted down the line, 'Quick, sir, there's a polar bear out. It's the male and we've no idea where he is.'

Sheer terror gripped me and my throat went dry. I managed to summon up enough breath to say, 'I'll come down the tunnel. Meet me outside the aquarium.'

This was the best idea I could think of on the spur of the moment; the aquarium was closest to my flat and most of the journey would be through a long tunnel, which an escaped animal would be unlikely to enter. The two tunnels – east and west – were the only connection between the south gardens of the zoo and the middle gardens. The Mappin terraces, which housed the bears, including the polar bears, were in the south gardens, and the aquarium was geographically the closest place to the terraces. I left the house wondering what I would do if I felt something warm and furry in front of me as I walked along.

My passage through the tunnel was uneventful but dark. The lights were very hazy and practically smothered by the smog, so I felt my way along the wall. I tried to walk with one hand in front of me, thinking, in some curious way, that this would protect me if I came across the polar bear; at least if I did feel the bear I'd know what it was! The thought of colliding with the beast in the dark sent shivers down my spine and I could feel the sweat trickle down inside my shirt.

I walked up the slope and out of the southern mouth of the western tunnel into a lighter sort of darkness and the swirling smog. The daylight was attempting to break through the smog and I could hear voices from the region of the aquarium doors. Having left the comparative safety of the tunnel I walked with both hands in front of me, calling out repeatedly as I went to identify where I was coming from. The smog was so thick that I didn't see the gathering of keepers until I was about ten feet away from them. They were keeping close together, very sensibly, but were obviously scared to move without some sort of plan of action. We thought the first move should be to go down the main walk in a line holding hands, with the end markers each carrying an escape net. When we saw the bear we would halt and hopefully smother it with a large escape net in the practised manner, so that he would enmesh himself. We had rehearsed this many a time and were extremely skilled at it.

We fanned out and dredged the main walk in the zoo with no result that we could see or feel. Suddenly, there was a yell from an end marker: 'Christ Almighty, I've just bumped into his arse. He's run away!'

We were about 200 yards from the bear dens when

I decided to stop hunting; it was too dangerous. We
would bait him back by scent. Slowly we felt our way
to the butcher's shop, which was a little way in the
opposite direction, and there we filled a bucket with lumps
of meat and another bucket with blood. We then felt
our way back to where we had been, poured blood and
meat on the ground and, as a group, walked to the polar
bear den, leaving a trail. We thought this was the only
trick available to us. The bear hadn't been fed since

the day before and would be hungry. Reaching the polar bears' den, we climbed onto the parapet above the open door and watched and waited in the swirling smog.

After what seemed like a lifetime, but was in fact only about twenty minutes, we heard a scuffling and snuffling and, quite suddenly in the gloom, we saw a huge male polar bear heading for his den. A few feet short he stopped. 'My God,' I whispered, 'he's scented us. Freeze, everybody.' We all froze and held our breath. The huge male stood on his hind legs and his mean little eyes stared at us for a few very long seconds at a distance which seemed like inches but was in fact about three yards. After these agonizing seconds he dropped onto all fours again and literally ran into his den after the food. With a big clang the bolts were shot home and we were able to come down off our perch. Very slowly in the gloom we found our way to the staff canteen, which by this time had opened for breakfast. We had to follow the fences and seats to get there. We were greeted by two bright canteen ladies. They had no idea what had been going on. One turned to me and said, 'Good morning, sir. You're up a bit early, aren't you? What's the matter, couldn't you sleep?'

Dead silence reigned and nobody said a word.

All I could think of to say was, 'Can we have six

coffees, please, with sugar?' Coffee never tasted better than the cup that morning.

The bear settled down happily with his mate and bred Brumas. So his escapade did him good!

TAIL
PIECE

CHOLMONDELEY GOES
WALKABOUT

HOWEVER VALUABLE A ZOO ANIMAL MAY BE – EVEN IF ITS
worth is measured in five figures – and however devoted
its keepers may be to it, the safety of the public always
takes priority in the very rare event of an escape. If it is
possible to take the animal alive by using capture nets,
traps and restraint boxes, then so much the better, but
otherwise tough decisions must be made. Such was the
fate of Cholmondeley the runaway chimp.

Cholmondeley was a chimp of considerable taste and
bonhomie; he enjoyed the minor vices and refinements of
man which, having been reared as an orphan by his
previous owner in Africa, he had acquired with no
undue effort. He got through twenty cigarettes a day,
enjoyed the odd bottle of stout and had kept his former
master company at table. He wore clothes and had
manners that were better than some. Unfortunately for
Cholmondeley, however, the imprint of man had made a

deeper impression on him than even these surface appearances indicated. The chimpanzee, who otherwise would have led an uncomplicated life in the trees, had been captured as an infant, and over the years he had almost ceased to identify with his species.

It was only when his master had to retire and found himself unable to keep him in domestic surroundings that Cholmondeley began to experience the disadvantages of his state. He was presented to the zoo to look after, and he truly hated the change. Little by little the chimpanzee's once happy spirit flagged and he became morose and sorely troubled. His health began to suffer and he showed signs of a variable temper.

In Cholmondeley we had a classic example of what can happen to an animal that is brought out of its natural habitat and taught to imitate a human. He acquires his trappings and eventually accepts the role imposed upon him as natural, losing all memory of his primitive past.

In such circumstances, it is easy to imagine the bewilderment and sense of indignity experienced by the chimp when stripped of his clothes and confined to the company of his far less sophisticated brethren. The routine of life in the den, pleasant though it might be to the average inmate, must have been deadly boring to one who had, in his time, been so high on the social ladder.

I have always held that a zoo environment, provided its standards approximate to those maintained by London Zoo, is, perhaps, preferable for most creatures to the hazards of an existence in the jungle, but there are exceptions, and Cholmondeley was one of them.

As the weeks went by it became obvious to us that Cholmondeley was the victim of a developing psychosis. At times he was subject to the most violent fits of passion and rage, but mostly he would sit unstirring in his den, refusing all food and drink and resistant to all attempts to dissipate his gloom. But one day he developed a new behaviour trait; his keeper found him stroking and squeezing his face, groaning and grimacing as he did so.

Called in to observe this latest manifestation of the animal's misery, I felt that it was caused by physical pain, very probably toothache, and decided to examine him, under anaesthesia, of course.

In those days, the only effective method of anaesthetizing an animal of Cholmondeley's size and type was by using a primitive form of gas chamber, a box into which chloroform vapour was piped through an inlet in the side. But although we managed to lure our patient into this device, he was by no means co-operative in the period that followed.

We turned on the gas and waited for him to lapse into unconsciousness; then we waited some more, for nothing appeared to be happening. The chimp was still on his feet and apparently unconcerned. Several more minutes passed, and still he failed to succumb. It seemed that the gas was having no effect at all. Taking a puzzled look at Cholmondeley, I found him staring back at me with some aplomb. With one hairy paw resting against the side of the box, and the other reflectively scratching his massive chest, he looked more at ease than he had done for months, and this only added to my bewilderment. Here, so it seemed, was a rare phenomenon, one that mocked all our past experience.

It was only when one of the team passed to the far side of the chimp and suddenly burst out laughing that we realized the reason for Cholmondeley's extraordinary resistance. His paw was being used for more than a prop; he had placed his index finger neatly over the mouth of the gas pipe.

Unfortunately, this incident was to be the very last example, bar one, of the intelligence and resourcefulness of this most remarkable animal, and the tricks that had once earned him a seat at the captain's table.

Cholmondeley's quarters in the sanny had a door that, because of its age and vulnerability, had been strengthened by steel. This he took quite literally in his stride. He left his den by the simple expedient of lifting the door frame from the ancient brickwork. Minutes later I got the warning 'Chimp At Large'. Of all the confused attempts to save Cholmondeley from the consequences of his actions, the one most near to success, and certainly the bravest, was that made by Bill Harwood, a zoo official of long service and experience, who was involved in so many of the zoo's activities that it was sometimes difficult to see where his responsibilities began and ended.

Bill behaved like a hero in the Cholmondeley affair. Regardless of risk, he gave chase to the angry chimp, caught up with him and, very sensibly, pressed on him a bottle of stout. Then, with supreme effrontery, and an incredible feat of persuasion, he somehow got him to sit down and join him in yet another pint.

For a moment it looked as though Bill's plan to detain the animal until help arrived to return him to his den was going to work. But then, as if sensing the net that was closing around him, Cholmondeley jumped angrily to his feet and made off again, more furious than ever. He was now beyond all efforts of salvation.

Cholmondeley reached Gloucester Gate in Regent's Park before our attempts at rapprochement were forced to a grinding halt. With dense traffic on the road outside, and scores of unsuspecting citizens enjoying the sunshine in the park, there was no leeway left for the unfortunate animal.

The society's marksman, whose talents were reserved for those few grim occasions when all soft options have failed, was reluctantly called upon, and arrived with rifle in hand.

He used just one bullet to despatch Cholmondeley, the chimp who, without wanting to, had become a menace to the species he had been taught to regard as his own. As we looked at his crumpled body, I don't think there was one of us that didn't feel a sense of guilt. When I remembered the way in which the animal had been humanized the incident smacked almost of fratricide, and I shall always look back on it as one of the most painful things to have happened during my career at the zoo.

It is true to say that, however well loved a zoo animal may be, the safety of people just has to come first.

TAIL PIECE

CHASED BY A GORILLA

MOST PEOPLE REACT TO THE WORD GORILLA. THEY think of a fearsome great beast with huge teeth, slavering and aggressive. Many years ago there was a film called *King Kong* which described a huge gorilla on the loose in New York which finished up climbing the Empire State Building. It was eventually machine-gunned from an aircraft but, before this, it had stampeded through the district, crushing houses and causing mayhem. The poor creature was supposed to have fallen in love with a beautiful young girl, whom he carried gently and carefully in his huge hands. This fantastic and ridiculous story achieved great fame throughout the world and was a major cause of gorillas getting a bad name, which they really don't deserve.

In the wild, gorillas keep themselves to themselves and are very shy, gentle creatures. Many people nowadays have seen zoologists like David Attenborough in close contact with families of gorillas, exchanging touches and other signs of mutual respect. In fact, the gorilla normally only gets roused in the presence of animals from which it wishes to protect its family. Gorillas are highly intelligent and interesting animals. They run very well-organized and disciplined lives, and, as such, it seems a pity to have them in captivity at all. Many zoos that hold them recognize their need for privacy, but are aware of the vital part they play in maintaining stocks of the species. A superb example of this is the gorilla collection in Jersey Zoo, masterminded by that remarkable author and creative scientific zoologist, the late Gerald Durrell.

Better zoos now regularly breed gorillas which is excellent, because man's predation in their natural habitats in Africa means that their territory is shrinking fast. Unfortunately, some hunters value a gorilla's hand as a souvenir, and you can imagine the slaughter required to supply this need.

For many years London Zoo had a very famous and well-known gorilla called Guy. He was huge and very gentle. Attempts were made to give him a wife, but he preferred his solitary existence; many people felt that his accommodation, although scrupulously clean and well cared for, was insufficient for such a huge and magnificent beast.

My first introduction to gorillas was with Guy and the head keeper who looked after him. The head keeper was a very wise man, with a great deal of knowledge about primates in general and the gorilla in particular. He had a very amicable relationship with Guy, and used to go into the cage and outer den to play with him, which astonished some of us.

When I was still fairly new to the zoo, I was taken by the monkey house head keeper into the mess to enjoy a ritual cup of tea while he explained his various skills with the primates and, in particular, with Guy the gorilla. When he told me that he used to play with this enormous animal, I was intrigued to know what he would do if the gorilla got cross or out of control.

'Aha,' he said and went to the corner of the mess and produced a short length of hosepipe, which fitted neatly into his pocket.

'What's that for?' I asked.

'You see,' he said, 'gorillas don't like snakes. If I want Guy to stop playing with me, I simply produce this little pieces of hosepipe, which he believes is a snake, and he backs off and leaves me alone, and I leave the den. Come back after feeding, sir, and I'll show you exactly what I mean.'

I returned after feeding, and the head keeper took me to Guy's cage, unlocked the door and entered. He walked up to Guy, who offered his hand, and they had a little parley and a bit of soft play in the inner house. Then the huge gorilla rushed into the outside den and made it clear he wanted to play outdoors with the head keeper. I arrived outside to watch the display. It went on for some time. I thought the play was getting a little rough, but the head keeper's pride was at stake. When it got to the stage that Guy had picked the head keeper up by his collar and the seat of his pants, and was literally dusting the floor with him – apparently in fun – I realized that the keeper was out of breath and getting more than he'd bargained for. As he passed on his second or third lap, he murmured in a loud stage whisper to me, 'Get the hosepipe; get the hosepipe.'

It dawned on me that the head keeper had forgotten to put the hosepipe in his pocket and couldn't stop the gorilla. I rushed back to the mess, collected the hosepipe, went to the outside cage and pointed it at the gorilla, who was now thoroughly dusting the floor with the head keeper. It worked like a charm: the gorilla dropped the head keeper and rushed back to the inside den and refused to come out. The head keeper came out through the outside cage, looking pink and shaken, and covered in dust and dirt. He rose to the occasion with great dignity, however, and turned to me and said, 'As you can see, sir, I have a perfect understanding with Guy!' This tale illustrates that gorillas can be quite playful and very gentle, but they don't know their own strength.

Some newly imported gorillas which had arrived at the zoo were transferred by me to the quarantine section for a health check. We needed to find out if they had any skin problems, check their intestinal contents, urine and so forth to see that they would be fit enough eventually to go out on exhibition. There were three silver-backed gorillas – a rather rare variety. The health checks were completed, except for an X-ray to make sure their lungs were clear of any problems. These gorillas were strangers and not particularly amicable, so it was decided that they would need to be sedated before they could be X-rayed. How to administer sedatives always poses a problem as most animals are a bit too wise to take anything that's smelly or tastes bitter.

The gorilla is a very intelligent animal and so we were stumped as to which sedative we could give him in his favourite orange juice without him tasting or smelling it. We had been sent a sample of a new drug from a famous continental manufacturer, which came with glowing references, saying that it was the ideal oral sedative for suspicious animals. It certainly seemed to be: it was colourless, just like water, had no taste when placed on the lips and no smell. We all thought it would do the trick. We mixed up the appropriate dose in fresh orange juice, and the mug was handed to the animal, who sat patiently wondering what on earth was going on.

It was an amusing sight to see us all nonchalantly waiting in the corridor for the drug to start working. First of all the gorilla eyed the drink with the greatest suspicion; he licked the glass and put the tip of his tongue into the liquid, he looked at us for a long time and held the glass very firmly, but he didn't drink. After a period of very careful thinking and consideration, the gorilla decided it was just a nice glass of orange juice and so he drained the glass to the last drop.

We now waited. We stared at the gorilla, and the gorilla stared at us. Five, ten, fifteen, twenty minutes went past with no apparent effect, except that the gorilla was obviously becoming amused by us staring at him and doing nothing. He kept jumping around the cage to see if he could do something to amuse us or make us go away. After twenty-five minutes, when we were giving up all hope of the drug ever working, the sitting gorilla suddenly shut his eyes very tight, rocked to and fro and fell over backwards with his arms splayed out. He was, to all intents and purposes, unconscious. Although it had taken all this time, we were very relieved that at last the drug had had the desired effect.

I always made it my responsibility to enter the animal den first to make quite sure the patient was unconscious and suitable for my staff to handle. My security lay in a rope tied round my waist, which was paid out as I went in, with strict instructions from me that if I said, 'Pull,' they were to pull as if their lives depended on it. I reminded them that mine probably might! So into the cage with the – hopefully – unconscious animal I went. I tested all the gorilla's reflexes, and he was apparently totally unconscious. I asked two keepers in, who lifted him up and took him down the corridor to the X-ray department, looking like a drunken sailor with his feet trailing behind. On a visit to Russia I had seen many vodka-soaked Russians being treated the same way by the police.

The little procession got about ten yards up the corridor, close to the X-ray department, watched confidently by me, when suddenly the gorilla stood up and threw the two keepers aside. He turned round and faced me; he was obviously very angry and extremely wide awake. I will never forget that sight; he raised his arms up, opened his jaws wide to reveal the scarlet inside of his mouth and rows of gleaming huge teeth and roared as he ran towards me.

For a fraction of a second I froze and didn't know what to do. I looked round desperately for a weapon with which to defend myself and saw an ordinary broom with a bristle head. In sheer desperation, I picked this up and held it out in front of me like a prickly bayonet or lance. How ridiculous!

The gorilla ran into it with full force and knocked me backwards onto my bottom. To my astonishment, the deflection off the broomhead flipped him back into the den he'd come from and, even more astonishingly, he fell down totally unconscious again. I went in and tested his reflexes and apparently he was out for the count once again. I abandoned the attempted X-ray, locked the door and went into the mess to discuss the problem. It was quite clear that the new drug, however wonderful it claimed to be, was totally unreliable if the animal could return to consciousness for long enough to create havoc and then fall back to sleep again. This gave no security to the people working in a zoo hospital.

In a few days we reassessed the situation and used drugs that we could depend on, by injection, having enticed the gorilla into a suitable restraint cage that allowed us to inject him. The X-ray routine went off perfectly satisfactorily and we were able to report upon it. We discovered that not only this one, but the other two gorillas all had respiratory tuberculosis, which they must have caught as youngsters in the collecting centres in Africa.

Some years before, we had developed a tuberculosis treatment and cure after a savage outbreak of TB in the collection, and we were able to administer this to the gorillas, who made a straightforward recovery.

This breakthrough was made after discovering that some keepers and animals had become infected with tuberculosis. The infected keepers were given sick leave and keeper protection barrier methods were introduced for the remaining staff, including mass testing and vaccination. However, amongst the collection of animals at the zoo the disease seemed to be spreading.

Where applicable, tuberculosis testing using tuberculin was carried out. Some animals were found to have human TB – presumably from infected keepers or zoo visitors – and others had bovine TB – probably from the sort of knacker meat they were being fed, which, due to government policy regarding the slaughter of farm animals, came from undisclosed sources. The normal policy for such an outbreak among domestic cattle would be to slaughter them, but with a zoological collection this was too drastic and emotive a measure.

At this time I met a French doctor called André Gremeaux, who was director of research for a commercial drug company. He was interested in our problem because his company was at the time involved in developing a drug aimed at treating TB in man. It worked on the principle of two drugs. The first dissolved the fatty envelope that protects the bacterium, and the second one killed the bacterium itself. He told me of a new combination of chemicals that were proving more effective than older treatments, and we decided that the zoo outbreak was the ideal opportunity to try out this magical combination.

We had 250 rhesus monkeys housed in Monkey Hill, and to our horror they had become infected; we built them a huge open-air sanatorium yard so they could be caged outside in the fresh air. All the infected primates were then treated with Gremeaux's formulation, and the combination of clinical isolation and his new therapeutics stopped the outbreak magnificently.

After this success, I presented a paper on the subject to one of the society's monthly meetings. In later years, therapeutic treatment of Third World populations severely affected by TB was introduced by medicated drinking water – all based on our experience in the zoo. More sophisticated treatments using antibiotics were later developed, but our work at the zoo was still something of an unsung breakthrough in the war against TB.

Some of my worst nightmares are those when I wake up yelling having been chased down an imaginary corridor by a gorilla. Are you surprised?

TAIL PIECE

ESCAPE

THE DAY STARTED OFF QUITE NORMALLY IN LONDON Zoo. I rose and visited the hospital as part of my early morning rounds at 8 a.m.

That morning's sick round and visitations seemed to be running to an ordinary pattern, and the efficient hospital staff were getting on with their duties and assessment of thirty or more diets for the sick and ailing that were at that time in their charge in the hospital. The hospital round complete, I visited my secretaries to find out what the rest of the day had in store for me. The next appointment was a meeting at 11 a.m. at the British Veterinary Association headquarters in London, so I prepared to leave.

As I left I said, 'Anything at all, telephone me at BVA headquarters and, if necessary, I'll come back straight away.' They looked at me rather wearily; they knew perfectly well what the instructions would be, but they tolerated me giving them the same ones every time I left. None of us realized that on this day they would be vital.

The meeting at BVA headquarters progressed monotonously, and my attention drifted from time to time and I longed to be anywhere else in the world than sitting at a table in a stuffy room in Mansfield Street. Suddenly the meeting-room door flew open and one of the association's staff came rushing into the room and straight up to me, bringing the meeting to an abrupt standstill. In a nervous voice that everybody could hear, he called out, 'Quick, there are two bears escaped from London Zoo.' The impact on the meeting was dramatic, but not as dramatic as my exit. I arrived at the zoo six minutes later, dashing through the swing doors of the hospital, which were left crashing to and fro with the speed of my entry. I had instructed my efficient secretary to have ready my capture rifle and pistol – the flying syringe – which I thought I would need.

The idea for the flying syringe came to me after I met Red Palmer of the Palmer Chemical Corporation in the USA. He had invented a rifle that was successfully used in the US for capturing suspect rabid dogs. I modified the syringe inside the dart of his rifle so that it held a cocktail of drugs – an immediate paralytic with a quick-acting tranquillizer and an anaesthetic. This meant that animals could at last be captured humanely, without the use of outdated nets and restraining boxes, and the drugs ensured the animal would have no memory of the event and would therefore suffer minimal stress.

This giant leap forward in medical management of wild animals, both in the wild and in zoos, became an international weapon. I made the technique available to all vets by describing it in the *Veterinary Record*, and it became colloquially known as the 'flying syringe'.

We then went on to develop a short-range weapon from a modified Webley air pistol.

I stayed long enough to pick up the information that the bears were roaming around somewhere inside the Mappin terraces, which are large curved terraces housing a variety of animals including bears. There were thousands of people walking around, and my fear was that the bears would escape from the terraces and attack the public. Capture nets had been placed all around the gardens, but the thought of trying to catch bears that had already mixed with the public caused me to quail.

I rushed out onto the lower of the Mappin terraces and, to my astonishment, saw three silent, motionless zoo officials.

I rushed up to the first one and shouted, 'What have you done?'

'I', he said portentously, 'have locked my office door.'

I had no time to think how stupid this remark was, and simply repeated the question to the second official, who was staring fixedly at the terraces, as if he could magic the bears safely back into their dens. 'What have you done?'

Total silence. The man seemed struck dumb by the potential horrors of the situation.

I turned to the third official and screamed, 'What have you done?'

'I', said he, 'have sounded the hooter.' How he thought sounding the hooter, which was normally used for clearing the gardens for other purposes, would assist us I do not know.

Realizing that time was being wasted, I rapidly roped off the terraces with the assistance of the keeper staff, and then entered the inside of the Mappin terraces, where darkness prevailed. By the light of a single bulb I saw a gathering of four keepers and heard a distant voice shouting, 'Where's Jonesy? Where's Jonesy? Where's Jonesy? I want Jonesy.' This was the agonized cry of a head keeper who knew there were two bears wandering about somewhere but didn't know quite where.

Inside the façade of the terraces there are, of course, access corridors that go round the whole semi-circle of outer walkways. These access corridors are necessary for feeding, cleaning and general hygiene.

At last I met up with the head keeper – a fine man but

now understandably terrified – and we devised a plan of action. I decided that we would all go up some narrow back stairs onto the second inside terrace, where it was most likely that we would find and capture the bears. The narrow rickety wooden stairs were only wide enough for one man, so we climbed in single file up the steps and emerged onto the concrete floor of the inner terrace. Because of the curving nature of the building it

was impossible to see more than ten yards ahead, as the rest of the terrace disappeared out of sight round the bend. There was no way we could see whether the bears were close so, in what seemed like pure pantomime, I told the men to arm themselves with steel dustbin lids from an array of dustbins which were awaiting the day's rubbish. This they did, and I told them to walk behind me as we went down the terrace.

I was prepared to shoot a drugged dart into the bears if I had time, but until I saw them I really didn't know what might happen. As we rounded the gentle bend of the terrace, we were devastated to see two large American black bears heading fast in our direction. Imagine it, six or seven men armed with dustbin lids facing two rampaging bears. We immediately discovered we were not heroes! We turned tail and fled, and I have never heard such a clattering and banging as we tried to get down the narrow stairs carrying our dustbin lids. We all finished up at the bottom in a great heap, cursing. Trying to retain some dignity, I stood up, dusted myself down and said, 'Well, we must try something else.'

In the end we put meat and bait in the outside walkways of the bears' dens, and eventually they walked back to get the meat when they got hungry. The keepers rammed-to the inner doors of the terraces and locked the bears safely back inside their dens.

TAIL PIECE

It was revealing how, in the face of two approaching bears, heroism evaporated and was replaced by common sense.

6

SUNBATHING SUKIE

SUKIE WAS A DELIGHTFUL SOUTH AMERICAN SPIDER
monkey with long black hair, beautiful big limpid eyes, a
topknot on her head and simply splendid red lips, which
formed into a petite 'O' when she called out with her
whistling call whenever she saw me. Sukie taught me
more about animal heating than anyone else.

London Zoo was built in the early nineteenth century,
and it led the way in what was then good zoo management. Many zoos the world over, having learned from
London's mistakes, built more modern and better zoos,
forgetting where it mostly started, boasting of their new
structures and inclining to patronize London Zoo. They
all forgot the famous Sir Stamford Raffles, who
spearheaded the need for wild-animal conservation and
collecting and made the Zoological Society of London
the prime scientific zoological institution in the world.
It still ranks with the best, has perhaps the most

comprehensive library of its specific type in the world and nowadays has huge new scientific research establishments (some may say too many for the budget, but nonetheless impressive worldwide).

But – and it is a big but – as the years progressed the original buildings, now well over 100 years old, had faults. Heating was one of them. Contrary to popular belief, many wild animals don't want constant heat, and certainly not convected heat of the stuffy office variety. They want to choose their own environment as they do in the wild, by selecting sunshine or shade. One hundred years of traditional radiator and electric heating in the zoo was hard to combat, but something had to be done because the animals were being cooked by convected heating.

The animal houses all tended to be hot and stuffy during the winter. The lion and monkey houses were shut up at 4 p.m. when the keepers went home, leaving all the heating turned on and little ventilation. This gave overnight indoor temperatures of 80 degrees Fahrenheit, conducive to disease, fermenting bacteria and lethargic animals. At eight o'clock in the morning all the animals were put into their outside dens so the inside ones could be cleaned. Thus they went from 80 degrees Fahrenheit to very low outdoor temperatures. It's hardly a surprise that quite a lot of them got ill.

I was very concerned about this. I worked out that God had given them splendid fur coats and the ability to stand enormous diurnal temperature variations in the wild – they may live in 40 degrees centigrade during the heat of the day and endure temperatures below freezing at night – so why did we need to keep them heated at night? In the winter I therefore decided to turn down the heat in many of the houses where the animals could be expected to acclimatize; and I left the connecting slides open all night so the animals could please themselves whether they went in or out. As a result, they all grew thick fur coats and became active. Insults were hurled at me from all quarters for doing this, including from the so-called bleeding heart brigade of misguided do-gooders.

The keepers had a meeting one Christmas time to protest. Traditionally on Christmas Day the grounds are closed and the keeping staff come in for an hour in the morning to feed and clean out their animals. This particular Christmas Day there was a heavy layer of snow on the ground. An official press photographer was going round photographing animals who were looking around wondering where the public were. He and I both witnessed lions and tigers rushing about in the snow in their outside dens literally playing snowballs. I met the grumbling keepers, explained my reasoning and showed them the press photographs of the lions and tigers. It was enough to reassure them.

In the old sanatorium, before the new hospital was built, we tried a new style of direct infra-red heating using special low-potency infra-red lamps supplied by Philips Electrical. We hung one of these lamps over the corner of all the cages in the sanatorium, allowing the animals to seek warmth if they wanted it, but avoid it if they didn't. All other heating in the old sanatorium was turned off. It was an immediate success. We had proved the point that pool heating, not general air-convected heating, was the answer, just as we would go in and out of the shade on a hot sunny day.

The excessive heating conditions of the radiator pipes, electric elements and so forth in the monkey house was conducive to ill health and the spread of unpleasant diseases like tuberculosis. Most commonly, many of the animals suffered from diarrhoea and colic. Little Sukie was no exception, and it was sad to see her one day with her long arms and legs curled round her, plaintive and ill in the corner of her overheated den. I rushed her to the old sanatorium, where she went into a cage heated in one corner by an infra-red lamp. She immediately took to the pool heating and lay stretched out underneath the light, making a spontaneous recovery very quickly. When she had recovered,

of course, she went back on exhibition in her old cage in the monkey house.

A few days later she was found lying huddled up in a corner of the monkey house crying plaintively, so it was back to the sanatorium. This happened several times before I realized, when all the tests were negative, that Sukie loved the infra-red lamp and feigned illness to get back to it.

Subsequently, I fitted infra-red pool-heating lamps in

the monkey house and rubber flaps to the outside dens with no slides at all. This enabled the monkeys to go in and out as they wished, day or night. All other heating was turned off, and in no time at all the monkeys were much happier, sitting in the 'sun' if required. The disease level among them dropped dramatically and their coats and tails improved, as did their activity.

The keepers were unhappy at the new vet interfering with the traditional system that had been in force for 100 years or more. Their protests were supposedly about animal welfare and my alleged cruelty to the animals by turning off the heat. In the end, after long discussions, I discovered it was the keepers who were cold, not the animals!

Pool heating, which is now so commonplace in zoos, was another first for London Zoo. Other zoos tried under-floor heating, but most zoos now heat with infra-red lamps where necessary.

GOLDIE THE EAGLE

THE MAGNIFICENT GOLDEN EAGLE IS RARE IN THE British Isles because of prevalent egg theft, and the few pairs that exist live in the distant highlands in Scotland. They are, of course, heavily protected.

Very few zoos have a Scottish golden eagle captive in their collection. This is just as well, because like many other broad and wide-winged birds of any size, their feeding and hunting behaviour depends on soaring to great heights. That, coupled with their incredible panoramic vision, enables them to pounce on unsuspecting prey. In the wild they will take lambs, large rabbits, hares and other birds as well. They would, should they exist in those parts, certainly take a dog. It seems obvious to me that birds that have a large wingspan of six or eight feet or more, which soar and fly long distances with powerful strokes, should have the space to do so and, ideally, not be housed in zoos at all. One like-minded zoo owner

I know made his aviaries horizontal, rather than tall and narrow, thus enabling the birds to fly quite long distances and swoop, turn and perform modest aerial aerobatics.

Goldie's aviary was a tall domed structure, built in Victorian times, when the mores of keeping such birds captive was not so avidly discussed. Those were the days when the zoo was first opened in the early nineteenth century, and spiked poles with buns on were sold to people so they could tease the bears in the bear pit. Those days, thank goodness, have gone. There is good and bad in all walks of life, and there are certainly good and bad zoos. Present legislation is getting rid of the worst offenders, but much still needs to be done.

Goldie had, I believe, been rescued from a trap that had been laid to catch a fox and baited with a dead rabbit. Goldie apparently swooped on the rabbit and was held firm by the trap. He was discovered by a poacher, who wisely threw his coat over him, collected him up and, having released his leg, took him to the local police. From there London Zoo was alerted, and the hospital medical facilities received Goldie and repaired the injured leg so that he could be put on exhibition. It would have been difficult for him to survive in the wild because the grip in his injured leg was insufficient to catch and hold prey and he would have lost his territory. Goldie was installed in the tall aviary, where he could land on a tree stump to eat his meat, and was much admired by everyone.

One day Goldie escaped during the routine cleaning of the cage. He had become very tame and enjoyed being hosed down, standing on his upturned log. Suddenly, and for no apparent reason, when the keeper's back was turned, he flapped through the open door and away.

Goldie zoomed far and wide around Regent's Park, and the zoo phones rang constantly with reports of the latest sightings. He would perch on the top of tall trees within a mile radius of the zoo, scaring the living daylights out of his other feathered friends who lived there – he was, of course, huge compared to them. Because he was unused to long-distance flying or soaring, his wing strength had been diminished by years in captivity, so he flew around in short flights from tree to tree.

As darkness fell we realized that he would roost soon and that there was no point in trying to do much more. Goldie chose a tree a quarter of a mile from the cage to roost for the night. Consultations between the curator of birds and others decided that the assistance of the Fire Brigade should be sought because, having perched and roosted, Goldie would probably be sleepy and might not fly away. Capture was in everyone's mind – except Goldie's!

The Fire Brigade turned up under the trees with their floodlights and long ladders, which were gently erected, to be within grabbing distance of Goldie. Intrepid firemen mounted the ladder complete with nets, and floodlights illuminated the scene. The press, of course, had a field day. Every time a fireman got to within two feet of the bird, Goldie would hop an infuriating two feet further out of reach. The ladder would be turned a little and the two-foot hop would be repeated. After a while it was considered that enough time had been spent by the brigade and the capture would have to wait until the next day.

The next day a close watch was kept on Goldie, who didn't move a great deal from his perch. He seemed to like the tree he had chosen.

During the day he made two more swoops to the ground to try and capture dogs being walked by innocent passers-by who were astonished, to say the least. Luckily for the dogs, Goldie couldn't grip because of his injured foot and his dive was clumsy enough for the animal to be able to escape.

Two days and nights passed, by which time Goldie must have been hungry. A simplistic trap was arranged, with two keepers sitting with a net baited with succulent red meat. For hours Goldie kept looking down at the meat, trying to fight his hunger and refusing to budge from the tree. Then, quite suddenly and without warning, he swooped down onto the meat, and the keepers flung the capture net over him and gathered him up in a blanket.

Goldie was taken back to his accommodation, where he received a lot of fuss and loving care from the keepers. He even had a special meal that evening, which he'll probably never forget: a whole rabbit. He was happier inside his cage than he had been flying around, undoubtedly because he associated himself more with man than with other birds.

Goldie eventually found a home in a specialist aviary, which was horizontal and so had plenty of space for him to fly.

83

SNAKES ALIVE

MANY PEOPLE HAVE A PHOBIA OF SNAKES AND INSECTS. During my work at London Zoo I discovered that snakes are very clean animals and that even the poisonous ones are normally docile, just so long as they are well fed and looked after. They do not enjoy the cold weather and warmth increases their activity. If they are hungry they also tend to be a bit less docile.

Most indoor exhibits of snakes are fairly static. The zoo built an outdoor snake exhibit, which was oval in shape and about twenty feet long and fifteen feet wide. There was a mound in the middle, which was covered with plants. The protecting wall was three feet above ground level with an inward-curved ledge to stop the snakes escaping, but sank to three feet below ground level to act as a dyke. This produced a moat of water around the island that was about six inches deep and a foot wide, and in it grew a variety of aquatic plants.

Both on the island and in the water there was a selection of water-loving non-poisonous snakes, so the public could observe sixty or seventy different species in safety. It was a very popular exhibit and visitors were always surprised to see brightly coloured snakes swimming about in water; most people don't realize that snakes are very good swimmers. They could also watch snakes basking in the sunshine on the rocks that peeped out of the mound.

We had little trouble with the exhibit, except during school holidays, when crowds of excited children would gather around the snake pit, showing either joy or horror when they saw the snakes. Occasionally the more foolish among them would stand on the parapet. Most small boys – I remember so well when I was that age – have a sort of hunting instinct, which shows itself in their wish to catch and mount butterflies or to go fishing with a jam jar, rod, piece of string and a bent pin to catch tiddlers.

There was one particular schoolboy who decided on a game of oneupmanship with his peers. They were all collecting marbles, tiddlers, butterflies or insects, so this little boy decided that he would have nothing less than a snake for a pet, and he aimed to get one from London Zoo.

He had obviously done some site investigation, and he

knew that, at night, the zoo was patrolled by night-watchmen who clocked in at certain points on their rounds to establish where they were should there be an incident or any intruders. You could easily gain access to the gardens, either over the front entrance turnstiles or over one of the fences directly from Regent's Park.

The schoolboy, who was about twelve years of age, thought his campaign out very carefully. He had visited the zoo several times with school parties and noted the watchman's clock-on points for his future escapade. We found out afterwards that he had equipped himself with a fishing net. He cycled to the zoo in the early hours of the morning and climbed over the turnstiles of the main gate. From here he only had a few yards to go to get to the open-air snake collection. He waited in the shadows for the nightwatchman to clock in at the nearest point and go on his way. All he had to do then was dash to the snake mound and the water course surrounding it and, using his fishing net, scoop up a snake. His plan worked very well.

Imagine a young lad of about twelve cycling home with a fishing net containing a surprised, damp snake. Luckily for the snake, the boy lived only half a mile away from the zoo. He was able to reach home, park his bicycle, get back into the house and go up to his room without waking his parents. However, he was then puzzled as to what to do with the snake. Should he leave it in the net on the floor? Should he put it in a drawer with his clothes? Should he put it in his school satchel? Should he put it in the bed with him? He decided against all these options as a master plan came into his mind. Having carried out his plan and seen the snake safely tucked away, he retired to bed and went to sleep.

Our sleeping thief was woken the next morning by the terrible screams and shouts of his agitated mother. The racket she created brought the whole family running, including the father.

'What on earth's the matter?' he said, trying to comfort his wife.

'Look,' she said. 'Look, look, look! Look in the toilet.'

Father walked over to the toilet and, sure enough, in the lavatory bowl, happily swimming around, was a coiled-up snake. Pandemonium ensued, voices were raised and panic set in as people flew about the place. The father marched into his son's room where he found the boy pretending to be sound asleep buried underneath the blankets.

He woke him up abruptly and said, 'What do you know about the snake in the lavatory?'

'Well, Dad . . .'

His father exploded. 'This had better be good. Your mother's frightened out of her wits, your brothers and sisters scared to death and I'm not pleased, so you'd better have a pretty good story.'

'Dad, all my friends at school collect all sorts of animals and they're always boasting about them and taunting me because I don't have a pet. I don't have a rabbit or a guinea pig or a hamster. I don't have fish or insects. I don't have any pets of any sort.'

'No!' exclaimed his father. 'Because this house is too small for them.'

The boy confessed to his father how he had planned to take the snake from the zoo in the middle of the night.

His father didn't know whether to roar with laughter or get cross. He was at a loss how to handle the situation.

'I'll give you five minutes to get that thing out of the lavatory and make it secure somehow, then you can telephone the zoo and tell them what you have done.'

The boy quickly scooped the snake out of the lavatory and popped it back in his net, keeping it captive with the pole turned over to shut the top of the net. The snake was then returned to the zoo, who were grateful, as they hadn't even noticed it had gone!

We all took a liking to this boy and his cheek. Such was his affinity for snakes that we gave him a Saturday job in the reptile house. He was brilliant at it. I saw in this boy a great kinship and bond with the animals, an essential quality for any veterinary surgeon. He so wanted to be a zookeeper, but his parents encouraged him into further education instead. A pity, as he was a natural.

ZEBRA CROSSING

IT IS A BIT OF A MYSTERY HOW THE HORSE SPREAD throughout the world. There are wild horses in Mongolia, supposedly the origin of the present equine species; they spread to Africa, and from Africa to the Americas, presumably before the oceans separated the land masses. One of the African wild horses is the zebra, a most handsome creature, well known by all stand-up comics as a horse in pyjamas! Different varieties of zebra have different striping to identify them.

London Zoo had a zebra stallion and two zebra mares who were kept in four roomy inside dens, which allowed for manoeuvring, moving, cleaning and separation. These dens had sliding doors, which were operated from the safety of the corridors outside. The zebras had two outside paddocks, which were inter-communicating. The first was accessed from the end den, and the second paddock was separated from the first by a high chain-link fence and field-gate. It could be used for separating the stallion from the mares so they could exercise outside without being bothered by the male.

Zebras are very tough. They are rarely tamed and put to use by man since they remain completely independent in their wild herds. They have feet like iron, and even in captivity they are never shod because it is almost impossible to cut their hooves. If their hooves overgrow in captivity, the only way to cut them back is to anaesthetize the animal and use a Black & Decker saw! They are far too hard to cut with any ordinary blacksmiths' tools. As to zebras' nature, well, they can only be described as mean as hell, particularly the stallions, who aggressively protect their females and have an inborn distrust and dislike of man. They seem to be athletic gymnasts, too. I think the zebra is the only animal that can jump up into the air off all four feet, kick you with each one in turn before it lands again and throw in a bite for good measure.

One of the two mares developed a huge shoulder abscess, and she needed an anaesthetic and an operation to put the matter right. This all had to be done in her den because at the time we had no proper hospital facilities.

The den was suitably prepared and the other zebras –

the mare and stallion – were put in the two outside paddocks so they wouldn't interfere. We darted the mare with a flying syringe, rendering her unconscious; she was connected to anaesthetic apparatus to keep her unconscious, and we were ready to operate. The operation to open up the abscess and fix drainage tubes to it took about three hours. After this, we left her peacefully sleeping off her anaesthetic while the other two zebras remained in the outside paddock.

At about ten o'clock I checked that the patient was standing and looking well and had drunk and eaten, then I let the other mare into one of the inside dens so that she could keep her chum company. All the slides were then shut and bolted, and I returned to my flat to prepare for bed.

At midnight the internal telephone rang and I took a call from the keepers' lodge, where the bachelor keepers stayed in order to have all-night staff on the premises. A very agitated young keeper told me there was a terrible noise coming from the zebra enclosure. I told the man to stand by the phone in case I wanted him, and I left my flat and drove through the zoo to the zebra house.

As I got out of the car the noise the stallion was making was absolutely horrendous. Both the mares were getting distressed, particularly the one recovering from the operation, and I decided that somehow or other the stallion had to be let into the end box so he could settle down and not keep the mares awake.

I worked out that if I opened the sliding door connecting paddock one and walked over to the gate that secured the stallion in paddock two, I would be able to quietly undo the chain safeguarding the gate, having released it from its padlock, and retreat to safety before the zebra realized his route to the end den was clear. This plan seemed fairly simple, but when I got to the gate the stallion was raging up and down, screaming blue murder, and he kept kicking out and attacking anything he could get near. He took a particular dislike to me and started to attack the gate.

I decided that the only thing to do was quietly undo the chain from the gate in the semi-darkness while the stallion was rushing round the paddock on the other side. This I did, and I hit the stallion smartly on the nose with the free end of the chain as he rushed at the gate. This not only stopped him in his tracks, but made him back away for a few vital seconds.

I flung the gate open and ran like the wind to the security of the inside den and the corridor. As I ran, I heard him screaming after me, and I realized that it was going to be touch and go whether I could cover the ground to the den before the zebra. I flew across the paddock, through the inside den and landed in the

corridor in time to fling the sliding door to with a crash. As it locked shut, the galloping zebra piled up against it with an awful crash. I had literally felt his breath on my neck.

It took me a few minutes to regain my stability. I then checked that the operated mare was all right and, having reassured myself that all was now quietening down, I locked up and returned to my flat and my bed.

At 8 a.m. the next morning I went down to visit the mare to see that she was all right; there I was greeted by a huddle of keepers, headed by the head keeper. They had found the stallion in the inside den and yet, when they'd left the night before, he was in the outside paddock. 'How?' said the head keeper. 'How, sir, did you get him in?'

I told them. The head keeper shook his head and said, 'You should never have done that, sir. He's dangerous.'

I thought, Now he tells me!

He said, 'Look, sir, watch this.' He took out a Rizla cigarette paper from his pocket and raised it, hidden, up the woodwork towards the steel bars of the stallion's den. The zebra stallion was standing with his bottom towards us. As the white of the paper came above the wooden part of the stall there was a bang, like a gun going off. I never saw the stallion move. The keeper showed me the two pieces of Rizla paper. He had placed them just inside the bars and the stallion had kicked with such force and accuracy that it had split the paper in two. This impressed me no end!

'I did say that you shouldn't have let him in, didn't I, sir?' the head keeper repeated.

This experience taught me a lot about zebras; it also taught me that you should never cross a zebra, or meet a zebra crossing!

The mare recovered completely and later gave us a foal by the angry stallion. They called him Oliver!

TAIL PIECE

10

WHEN SABRE'S HEART STOPPED BEATING

FROM THE TIPS OF HIS SHAPELY PAWS TO THE CROWN OF his velvet-smooth head, Sabre the puma was an animal of dramatic beauty. His walk was a rhythmic swagger, his whiskers were cavalier, and his every action proclaimed his noble savagery. But the most striking of all his physical attributes was his calculating arrogant gaze. Cool and yellow, his oriental eyes surveyed you through the silky fringe of their long lashes, sizing you up and down, stripping you of pretensions. If ever there was a cool cat it was Sabre.

He had come to us from an unexpected source: the Royal Canadian Air Force. There he had served as mascot to the Cougar Squadron – the cougar being the North American name for the puma – until his owners had gone home, trusting him to the zoo's care in the early Fifties.

Sabre was over thirteen years old, but was as active and resilient as a cub, though without a cub's frivolity.

Weighing 260 pounds, he had one of the cleanest medical records in the zoo. His skin seemed to glow with health and vitality.

Then, quite unexpectedly, something happened to shatter the confident calm of this aloof aristocrat. His keeper reported that he was showing signs of stress and appeared to have a wound on his flank. After a look at Sabre, we came to the conclusion that he had developed a deep-tissue ulcer, and he was aggravating it by his persistent licking.

In a man, such a condition could be treated easily and cured. With an animal the process is far more complex. You cannot swathe a puma in bandages or smooth healing ointments over its smarting wound. In fact, one can do very little without sustaining damage. A sick wild animal is a very poor patient, and anyone who approaches it too closely is at risk. Slight though the trouble might be, we therefore had no option but to operate, which meant using restraints and anaesthesia, as would have applied to the most major and complicated of operations.

One of the most frustrating and least appreciated aspects of operating on a wild animal is the amount of time and effort that's used before you can even bring the animal to the table. The patient can't possibly explain to you how it feels and, as *you* can't explain to *it* the various steps you intend to take for its relief, you have to implement a programme that involves a whole series of separate manoeuvres – each of them vital – that demand an expenditure of resources and time that often seems disproportionate to the scale of the problem.

By this time our animal hospital had been constructed to replace the sanny. Designed by the Zoological Society's architect, the hospital compared favourably with any contemporary institution for the medical care of humans. It could cater for approximately 98 per cent of the zoo's inmates – mammalian, avian and reptilian – and had facilities for the most complex operations and the most advanced methods of treatment.

There were twenty-one roomy dens lining the corridors of the first floor and a recreation and exercise space as well.

Giraffes, elephants and hippos were obviously among the few patients who had to face their problems in their own surroundings, but the hospital was important even to these animals, their treatment being much improved by the experience acquired within its walls.

When we first started to dream of a replacement for the sanny, there was not proper transport available to move the animals from their cages. I redesigned the transport arrangements so that they could travel to the hospital in an electrically powered ambulance, specially equipped for their comfort and security. This was just one of the ways in which things altered for the better.

A sick animal can lose the will to live far more readily than man, who has the capacity to understand the reassurances of his doctors. To remedy this, we employed frequent changes of diet, fresh air and last, but by no means least, frequent visiting to reassure the animal and encourage its recovery. The sight of someone familiar – normally its keeper – often does more for a sick animal's morale than any amount of medicine. However, the innovation that brought the hospital the most prestige was its highly modern, fully equipped operating theatre.

To move our puma into the hospital without incident or injury, we first had to tranquillize and box him. Next, on the night before the operation, we had to administer sedation, in order to prevent his being disturbed by the unfamiliar surroundings. We didn't want him to lose confidence and therefore natural resistance. Then, on the morning of the day itself, we had to tranquillize him yet again for his trip to the operating table. This entailed his being transferred to the specially constructed restraint box; once this was done we were able to reach through the bars that formed its sides and restrain his back leg, injecting him with a light surgical anaesthetic. Thoroughly drowsy, our patient was removed from the box, lifted onto the stretcher, carried to the table and secured there. The critical stage of the pre-operational programme was then complete.

There were, however, further things to be done before we could commence the operation itself, and the team performed the functions with the smoothness born of practice.

A tube was inserted into the puma's trachea, which was connected to an anaesthetic machine to maintain anaesthesia. Next we fitted a drip to the puma's leg before turning to the site, which needed to be clipped and shaved and made secure against infection. Then we changed into sterilized surgical gowns and masks and applied sterile surgical drapes to the patient.

We did these things at speed, for time was not our ally. The animal had to be kept anaesthetized to protect the team from being attacked by their patient, but the anaesthesia could not be so deep as to interfere with the pace of the puma's subsequent recovery; the gas mixture therefore had to be adjusted constantly.

When we inspected the ulcer we found it was quite a monster. It was three inches wide and copiously suppurating, but we foresaw little difficulty in the task of removing it. Sabre appeared to be comfortable, and the operation was a purely routine affair. But then something happened that was completely unexpected: a cardiac arrest. The puma's heart had stopped beating due to hypoxia – insufficient oxygen.

Although gas anaesthesia is carefully monitored using a Boyles anaesthetic machine – the same as is used in NHS hospitals – some patients have a sudden heart 'shutdown', undoubtedly due to surgical shock, which is all to do with the conflict between the anaesthetic gases and the patient's adrenal system. The gases can sometimes cause a temporary shutdown of the adrenal gland, which supplies the adrenalin that is essential for the heart to function properly. Cardiac arrest can follow.

I swiftly injected the animal's heart with adrenalin, and then commenced external massage – a vigorous, thumping rib massage – to stimulate the motionless heart. Half a minute later, it resumed its natural beat and we could start again; the patient had stabilized.

By chance, one of Fleet Street's most respected photographers, Freddie Reed of the *Daily Mirror*, and his reporter colleague, Betty Tay, who specialized in stories about animals, were in the theatre at the zoo's invitation, at the time. The society had given them what was, in those days, the rare opportunity of recording a zoo operation, and, thanks to their wide experience, neither was the type to sensationalize what they saw. However, even Miss Tay was sufficiently impressed by the occasion to record, 'I watched as they massaged the animal's heart for thirty seconds. It seemed to me more like thirty minutes!' Only when the puma was being wheeled out of the theatre and into the hospital's convalescent ward did I realize that my face was running with sweat.

Although Sabre's operation was successful, wild animals aren't, on the whole, easy to treat – either by operation or medication. They do not have man's positive mental attitude, which helps us to recover from illness or an operation. Should the wild animal fall sick in his natural habitat, the chances are that he will die, either because he isn't fit enough to hunt for food, or because his illness leaves him easy prey to predators. He could be regarded as a burden to his herd or group, endangering their safety, and thus be abandoned or destroyed by them. Where a sick man can seek support from friends and family and even the state, the sick wild animal is everyone's enemy, and he knows it, and accordingly he is vulnerable.

The doctor is a symbol of recovery to the human patient; he is important for what he represents as much as what he does. But that symbolism is lost on an animal. We found that the best recourse when one of the animals in the zoo was ill was, except in the event of an emergency, to leave the patient to his own devices for the first twenty-four hours of his illness. Any attempt to intervene during this initial period could prove counterproductive. Medication is, after all, completely alien to a wild animal, and he will detect, and reject, any attempt to administer it, so it often causes more emotional upset than good.

At the zoo, we found that certain minor changes could aid an animal's recovery. These entailed firstly removing the animal from the environment where it became ill, and moving it to new and more acceptable surroundings. We would keep the patient in the quiet and warmth, and in semi-darkness, and feed him a selection of pleasant foods. In addition to this, the animal would be nursed by his keeper, the person most familiar with his day-to-day routine, whims and fancies; the keeper would initially prove a far better nurse than the veterinary surgeon. We considered the intervention of the vet as a last resort, and one to be called upon only when the animal's emotional status had been stabilized, and medical and surgical action could really be of benefit.

Sabre recovered quickly, and soon his hair grew back to cover his operation scar. He showed less stress and angst than any of us — but then, he was a tough puma.

TAIL PIECE

TALK OF THE DEVIL!

THE ANTIPODES, IN PARTICULAR AUSTRALIA, PROVIDE many zoological creatures of considerable interest, many of them marsupials. The best known, of course, is the curious kangaroo, with its ungainly hopping gait, its great speed, its powerful hind legs and small forearms and the pouch in front of its stomach for carrying its young to semi-maturity. When born, a baby kangaroo is only about an inch long and manages to creep up a licked path into the mother's pouch, where it attaches itself to a nipple. It remains there until it becomes a joey, old enough to hop out of the pouch, forage for itself, and hop back in if there's any trouble. What a handy arrangement. No prams!

Tasmania, the island just south of Australia, is home to some fascinating creatures and this tale concerns one, the Tasmanian devil. It is very aptly named; ask any member of the keeper staff of London Zoo, but it is still sad to think that because of man the Tasmanian devil is probably now extinct – but that is uncertain because they are difficult to find in the wild.

The Tasmanian devil is an ugly brute; it is roughly the size of a springer spaniel and has a dark, scruffy coat with evidence of striping. Its large head is equipped with masses of strong teeth in two large jaws. The neck and chest are also relatively large, but the rest of it tapers off miserably into a wretched little bottom with a hairless tail with a tuft on it. Not a pretty sight! It is a wild carnivore that kills and eats small mammals. It is rumoured that, to capture sheep, it is capable of chasing after one and biting its leg off to stop the sheep in its tracks so that it can then kill and devour it.

The savagery of this wretched creature is obvious when you approach it; it screams and snarls with rage for no particular reason and tends to attack on sight. It is difficult to catch and impossible to handle. You have to use nets or a flying syringe to capture it. Keepers would tell you that on no account must you get near it while it is conscious.

The Tasmanian devil is a hateful, uncompromising, smelly creature. In my view it is one of God's errors of judgement. This particular devil was in London Zoo in the Fifties and was safely tucked up for the night in its den – or so we presumed. No-one knew, of course, that during the early hours of Saturday night it had escaped. It had bitten through very thick, strong chain-link mesh, as if with a pair of bolt cutters. Imagine the fearsome strength of its jaws.

It was in Baker Street, London, close to Sherlock Holmes's flat, where the incident took place. The savage monstrous creature must have waddled off, going south-west across Regent's Park, probably passing Gloucester Terrace and other places of interest on its way until it entered the top of Baker Street. Here, presumably, it shuffled along the pavements, investigating dustbins and chasing cats. The nasty little creature must have crossed the Marylebone Road which, fortunately for it, was none too busy in the small hours of the night. Anybody seeing it dodging along could be forgiven for thinking it was some strange mongrel dog on a dustbin hunt. And so it was that the devil arrived outside a nightclub halfway down Baker Street. Any drunken reveller seeing it could well have believed he was seeing things. It was lucky for the revellers that it decided not to go down the stairs into the club.

One man emerging from the club, obviously full of the juice of the grape or the grain and tottering slightly, looked down and stared disbelievingly at the strange four-legged object by his car. Even with his somewhat blurred vision and unsteady stance, he realized it must be an escapee from London Zoo. There were no keepers and nets to be had. So what did the reveller do with this vicious, savage, aggressive beast? Send for help? Send for the police? Send for the RSPCA? Not at three o'clock in the morning! Unbelievably, he bent down and picked the vicious thing up by the scruff of its neck and hurled it, snarling with rage, into the back of his car, where it continued to rage and spit.

The man took no notice, got into his car, turned round

and drove up Baker Street to Regent's Park, where eventually he arrived outside the main doors of the zoo's head offices with his screaming passenger. He banged and clattered on the door until the resident housekeeper emerged, both surprised and angry at being disturbed. He looked in the back of the car and couldn't believe his eyes.

He rushed to the keepers' lodge, telling the driver to follow him in his car with the devil. The keepers emerged and were amazed by what they saw sitting in the back of the man's car. They rushed indoors for pole nets to take the animal captive without getting near it.

In a slurred voice the man said, 'What's the fuss?' Again he leaned into the back of the car, picked the thing up and plonked it, screaming with rage, into a pole net. He then challenged the keepers, 'What's up? Are you scared or something?'

Having disposed of the animal, he drove off into the night. The staggered keepers took the struggling Tasmanian devil back to where it came from and secured it in an inside house for the night.

The next day they saw the damage to the chain-link netting, by which time, of course, the press had been informed by somebody. They turned up and photographed the beast, the hole in the netting and the keepers describing the dreadful nature of the animal.

They revelled in the story of the tipsy nightclubber who, not knowing anything about the creature's nature, and being drunk, had known no fear and had grabbed it by the scruff of its neck in a way no keeper would ever dare to do.

The press wrote the reveller up as a hero and a brave man, conscious of his duties, but obviously he just didn't know what it was.

The Tasmanian devil died eventually of happy old age but could not be replaced. Funnily enough, no-one shed many tears, least of all the keepers!

TWO LIONS AND HITCHCOCK

WHEN I WAS APPOINTED CURATOR OF MAMMALS AND veterinary surgeon of London Zoo, I was combining the duties of two jobs. As curator I was responsible for staffing, animal accommodation, movement, breeding etc., as well as all my veterinary duties concerned with the health and well-being of the resident animals.

In 1957, after several years of my joint appointment, the workload was far too heavy for me to conduct either side of the job efficiently, and so I asked for the appointment to be split. This was done, and I became solely the veterinary officer – later the senior veterinary officer – so that I could concentrate on my scientific work to improve the health and welfare of the animals of both London and Whipsnade Zoos.

On one particular day I finished my morning rounds at the lion house. There, I joined the head keeper and his staff in their mess for morning tea.

The lion house had two parallel rows of outdoor and indoor cages. The outdoor cages were viewed from behind a barrier, and the indoor cages were viewed from terraces covered over by an arching roof. The space in between was roofed over and became the service corridor for both the outside and inside dens. Connecting the dens was an arched bridge whose sides were steel bars, like the top, and the floor was stout wooden planks covered with tarmac. Routinely each day the animals were shifted from one side to the other to allow for cleaning, feeding and other activities. In wet weather they could be in the indoor cages and in fine weather they could be in the outdoor cages. The system was guarded by sliding steel doors, padlocks and bolts.

Each house had a head keeper in charge of the staff and organization of the daily duties. The lion house was considered to be the best, and the keepers were proud of their roles. The head keeper of the lion house at this time was a magnificent man called Alf Hitchcock, known to everyone as Alf. Alf ruled his empire extremely well as supreme monarch. He was a magnificent, big, burly gruff cockney, with grizzled grey hair. He didn't suffer fools gladly and, I believe, barely tolerated me. But he loved his job and was brilliant at it. He used to lose his temper from time to time, as most of us do, but when Alf lost his temper he behaved like one of the lions, commanding a vast collection of swear words and blistering vituperation of awe-inspiring proportions.

We were gathered together over our mugs of tea, talking generalities, when there was a terrific crash and a roar. We all dropped our tea and rushed out into the service corridor. What we saw was desperately alarming. Four bridges away, standing in the dimly lit service corridor with their eyes shining pale green like headlamps, were a huge lion and his lioness mate. Afterwards, we found that the wooden bridges from the outside to the inside cages had rotted; they were, after all, 100 years old. It seems that on this occasion the weight of the pair crossing the bridge finally proved too much for the floor, which collapsed and deposited them onto the floor of the service corridor, producing the crash we had heard.

We all stood frozen in our tracks as the male and female advanced on us, growling fearsomely. There was no escape; we were trapped like Christians in the Coliseum in Rome. It was terrifying, and nobody moved as the lions advanced because we were frozen with fear. It seemed as though we were awaiting our fate.

Suddenly and explosively Alf Hitchcock decided he had had enough. He snatched up a besom broom and rushed at the pair, roaring his head off, threatening them with the broom and letting fly a prime example of his famous stream of invective. The expression on the lions' faces was hilarious; it was one of sheer terror. Obviously even lions could easily be intimidated by Alf in a temper!

They stopped in their tracks, but Alf didn't. He raged on towards them, waving his besom broom like a flail. He got to within feet of them, and we all held our breath. Then, with one final terrified look, the lions turned tail, fled down the service corridor to escape from Alf and hastily jumped back onto the broken bridge and rushed for the safety of the cage they had come from. It seems that lions will face anything except Alf Hitchcock!

Alf threw the bolts to secure them in their cage and then rejoined us. He was still furious. Not at us, but at the impertinence of the lions for interfering with his routine.

He snarled as he went into the keepers' mess, 'Where's my bloody cup of tea?'

Hitchcock the brave – Alf – retired soon after this event. He was one of the last of the one-house keepers, firmly rooted in tradition. I introduced a rota system for all the young keepers, so they could gain mixed experience, which would be more valuable to them when promotional opportunities came up. The old one-house staffing system meant they had to wait to fill 'dead men's shoes', and it had to be changed. But there will never be another Alf.

THE RAVEN WITH THE WOODEN LEG

THE BLACK RAVENS OF THE TOWER OF LONDON HAVE become famous. I am not quite sure how long they have been there, but I believe it's a century or more. In any case, they became, and are still, linked with the Tower in such a way that fable has it if the ravens ever leave the tower, or die out, the monarchy will fall. This is a most romantic piece of nonsense, but it's a great attraction to tourists, who are particularly keen to see the birds.

Ravens are fine big birds, heavily built and with powerful bills with which they attack their meat. They are mostly carnivorous and are treated as pets by the Tower guards or Beefeaters. The Beefeaters regard these ravens as their personal responsibility and each of the birds has a name and number. They are pampered, fed and looked after almost as well as the Crown Jewels themselves.

A tremendous panic started one day when a Beefeater spotted one of the ravens flopping about helplessly, flapping its great wings, extending them and trying to use them as support, and looking thoroughly miserable. He rushed to collect up the unfortunate bird, which by this time was not only frightened but angry and tried to bite him in the face. He rushed indoors with it, and he and his friends examined the raven, only to find, to their horror, that the lower half of the bird's right leg, including its foot, had been severed and was nowhere to be found. The poor flapping creature was unable to bear its weight on one leg and was in great danger of hurting itself further.

There was an urgent call to the London Zoo hospital, and the bird was rushed up for us to see. Immediate surgery was carried out on the stump to tidy it up, the bird having been given a very light anaesthetic.

As a temporary measure, we applied a firm tube of plastic over the stump and fixed it with elastic tape. This at least gave him balance, so that he was able to rest on the ground, although unable to perch. More needed to be done, so I rang up my colleagues at the Royal National Orthopaedic Hospital. A great friend of mine, Dr – now Professor – Scales, came down to assess the situation. Photographs of the left leg and foot were taken, and measurements of the remains of the stump very carefully made, giving a clear indication of what needed to be done if a prosthesis was to be made, which it was.

The bird was allowed to rest, and after a few days I drove it to the orthopaedic hospital in Stanmore. Here, a cast of the stump and the remains of the leg was made, and a matching cast of the good leg was taken to act as a guide. The bird returned to hospital for the time being while a prosthesis was constructed.

After a week, I was telephoned and told that the wooden leg was ready. A type of foot had been constructed, but without the flexibility of a normal one. It had a platform sole, upon which the bird could rest firmly, and which would give it total stability. With the prosthesis it should be able to march about, take short flights and land, but it would only be able to perch once it had learned how to rest one leg and grasp with the other foot.

We travelled up to the hospital again, where several prostheses had been made to make sure the fitting was exact. The bird had a light anaesthetic while the fitting took place, and when the raven regained consciousness it had a prosthetic right leg of the same height as its good leg. We liberated it onto the floor of the operating theatre. First of all it stood very still, not quite trusting the new leg. Then it took one or two short hops, using both legs at the same time. It was fascinating to see how quickly the bird gained confidence in itself. Soon it was actually striding round the room, not just hopping, all

the while making raucous noises, perhaps cries of joy at being able to walk about again.

The bird was returned to the Tower of London and it promptly mixed in with the other birds and continued to lead a normal life. Very few people realized it had a prosthesis at all, because the colour had been carefully patterned to match.

Quite by chance, the Beefeater who had particular charge of this bird was photographed by a professional photographer representing a firm of camera-film manufacturers. The wonderful photograph of the two heads — the raven and the Beefeater very close together and looking at each other — was enlarged to form an advertisement poster.

AN ELEPHANT NEVER
FORGETS

LONDON ZOO NOW HAS A VERY SMART NEW ELEPHANT enclosure, designed specifically and described as representing a circle of elephants' bottoms, or so one of the architects concerned declared! The old elephant enclosure in the middle gardens was, in fact, a conversion from two very large bomb-proof shelters constructed early in the Second World War to protect staff during a bombing raid. It was decided after the war to turn this accommodation into a temporary elephant shelter until such time as a custom-built house was built. Both the shelters were underground and made of reinforced concrete, with two doors opening onto a sloping walkway up to an open-air compound. These underground dens were very big for people, but none too big for elephants. The two concrete dens were connected at the back by a reinforced concrete tunnel, which opened into each den, allowing keepers to move from one den to another without being interfered with by the elephants.

The parade ground at the top of the slope from the dens was a large circular area, suitably sanded, with an artificial pool for the elephants to bathe in. Surrounding this circular paddock was a large U-section concrete dyke, the edge of which was surrounded on the inside by a row of spikes to stop the elephants balancing their feet on it. The outer wall was about three and a half feet high to allow the public to see but not get too close. The width of the dyke was eight to ten feet. In its way it was a very dramatic exhibition, with the elephants somewhat higher than the visiting public, emphasizing their huge size.

They were a popular exhibit. The public liked to lean across their part of the dyke, offering the elephants food – which we had to ban in the end – and the elephants would get as close to the edge of the paddock as they could and lean forwards, extending their trunks in order to take the titbits. Two elephants used the accommodation, an Indian cow elephant called Rusty and a great big African cow elephant called Diksie. Diksie was a formidable sight; she was quite huge, and had the vast flapping ears of the African elephant. She was also mean-tempered. Diksie became adept at balancing like a ballerina on her two front legs and leaning as far as her weight would allow to extend her trunk to the public. One of her tricks was to blow spit in the face of somebody who didn't have any food or didn't give her some. One day an angry lady hit Diksie's trunk with her umbrella, and, in one swift move, Diksie snatched the umbrella from her, crushed it to a pulp in her enormous jaws and swallowed it. The lady was extremely cross; she hammered on my flat door in fury, demanding a replacement umbrella!

It was Diksie's balancing trick that led her to disaster one day. The competition between the two elephants for food was great, particularly when there weren't many visitors at the zoo and titbits were in short supply; they would push each other sideways to get closer to the public. One day, while Diksie was doing her ballerina act on the perimeter of the paddock, Rusty nudged Diksie so hard that she fell with a crash into the dry dyke. There were screams of rage from both elephants and consternation from everybody near by. In the fall, Diksie tore her abdomen quite badly on the spikes of the dyke.

Getting an elephant out of a dry dyke is a problem. They weigh several tons and lifting them can be awkward and dangerous. This time, by building steps and stairs of tightly packed trusses of straw, we enabled Diksie to walk up and back out into the paddock. We agreed the best thing to do was to get her into the underground dens, where I could have a proper look at her wound, assess the damage and, if necessary, repair it.

After we got Diksie in her den we realized what a

close fit it was. Her shoulders were but a few inches from the ceiling and her huge bulk made it almost impossible for her to turn round. This unsuitable accommodation was eventually replaced.

I returned to the zoo hospital to get staff and equipment in order to sew the nasty gash in Diksie's abdomen. After using local anaesthetic, I inserted twenty heavy-duty sutures in the wound, which took about an hour and a half, crouched as I was in an uncomfortable position underneath Diksie. An elephant forms an attachment to a particular keeper and, while he is present, she will do anything he tells her. This bond is very valuable, but can be awkward when the keeper goes on leave. The presence of her keeper throughout the operation meant that Diksie remained calm and steady. It wasn't an ideal operating position for me to be in, but one gets used to it. The operation completed, I told the keeper about daily hygiene of the wound and arranged for my hospital superintendent to visit daily for check-ups.

After ten days, I returned with the hospital superintendent and, helped by the elephant's keeper, removed the sutures from the wound, which had healed and looked very healthy.

My office was not far from my flat in the middle gardens, and the walk took me past the elephant site. Walking home one day soon afterwards to have my lunch, I decided to call in and have a look at Diksie's wound. She was shut in the inside den, underground, because the keeper had gone off for lunch.

I opened the huge steel doors to Diksie's den and shut them behind me, a normal zoo precaution. Diksie was weaving to and fro, grunting, and she didn't seem to mind my presence too much. I walked round behind her into the area of the intercommunicating concrete tunnel. From this vantage point I moved underneath her to examine her wound. While I was doing this, I realized with horror that the elephant was slowly coming down on top of me, like a massive lift. She was trying to crush me up against one of the corners of the den and lie on me. I was terrified, and my one thought was of escape. Mercifully, I had positioned myself by the inter-communicating corridor and was able – just – to squeeze past her bulk into the end of the corridor. Diksie started screaming with rage, having lost her prey, and her angry trunk followed me up the corridor, forcing me to go as far as I could to get out of her reach.

To my dismay, the Indian elephant Rusty also lost her temper and joined in the fracas, presumably because she was frightened, and she tried to seek me out with her trunk from the other end of the tunnel. I stood very still, with both trunks touching my clothing in a frantic endeavour to get hold of me and drag me out. I have

never stood so rigidly still and terrified in all my life; I could feel sweat running down my body in every direction. The noise was sickeningly intense, with both elephants screaming in the confines of two concrete air-raid shelters.

Fortunately the noise was so great that the keepers came running from their lunch room. The two keepers attached to the elephants opened the doors and got the animals out immediately, up the slope and into the paddock. Then they shut the steel doors to prevent them coming back and came to rescue me. When they got me out I was faint with shock and fright, but I quickly recovered. The terror of those few moments lives with me to this day.

The keepers were angry with me for going in on my own without them as escorts, and they were quite right. They were wise and I was foolish. Obviously Diksie resented my operating on her and, after all, elephants never forget!

TAIL PIECE

I refused to countersign the plans for the new elephant house because there was no provision for a sky-hook attachment on which to fix a block and tackle for emergency lifting. Meanwhile, I left the zoo and moved to the university to teach. Diksie fell into her new dyke one day and broke a leg. She died. We never learn, do we?

15

CAESAR TO JERSEY

CAESAREAN SECTIONS THESE DAYS ARE COMMONPLACE when difficulties emerge that prevent the offspring being delivered naturally. The only difference in the Caesar I am talking about is that it was on a lioness.

The zoological world produces a lot of interesting people, quite a few eccentrics, a lot of scientists and some extremely good friends. One of the best friends I had was the late Gerald Durrell, the remarkable scientific eccentric who founded the Jersey Zoological Society and Trust. This remarkable man created what is probably one of the finest small zoos in the world, with one objective in view: conservation. He even repopulated an island, by replacing species that had long ago been decimated, to prove that it could be done.

At the time, I was living in a house in London, a few minutes away from the zoo. It was a Saturday morning, and I was busy in the garden pruning my roses when the telephone pealed. As I picked up the phone a voice said, 'Oliver, you are coming to Jersey. A car will collect you within half an hour; your ticket awaits you at the airport.' I recognized Gerald's voice and was a bit petulant because I had no idea what was going on.

'What on earth do you mean, Gerry? I'm busy pruning my roses; I have no intention of coming to Jersey.'

'A car will be with you in about twenty-five minutes. Your ticket is at the airport and you must come immediately. We have a serious problem.'

It turned out that one of his lionesses had been attempting to have cubs for two or three days and had gone to ground in her lair underneath the central island of the lion exhibition, a lair which subsequently proved to be large enough for a lion to walk into, but not tall enough for a man. If a man wanted to go in he had to crawl.

Sure enough, a hire car called for me and took me down to the zoo's hospital, where I collected all the apparatus I would require for a Caesarean. I also picked up the flying syringe, suitable drugs and special endotracheal tubes for that sized beast – these are tubes that go down the animal's windpipe and connect him directly to the gas anaesthetic apparatus. For windpipes the size of lions, giraffes or larger mammals, special enlarged tubes have to be made.

When I got to Customs at the airport with my luggage, I declared the fact that I had a weapon with me; this caused a mild furore and a rush of customs officials. I explained very carefully that it was a medical gun which fired darts containing drugs in order to anaesthetize animals. This was met with stony glares from all the officials; I'm sure they thought I was a terrorist about to blow up an aircraft. In the end, I was informed that the only way the weapon could go to Jersey with me was if it travelled in the pilot's cabin under his personal supervision. I readily agreed.

It originally took a long time to get the weapon into England from America because it sat in Customs for nearly 18 months. The reason for this was that it was classified as a guided missile, which had a military overtone. No matter how I tried, I couldn't persuade the Foreign Office that it wasn't a guided missile but a medical instrument.

We boarded the aircraft and set off – the baggage in the baggage locker and the capture pistol with the pilot. Arrival at Jersey airport caused some sensation; the press had already been alerted and a fair crowd was there. I waited in the customs hall to get my baggage, which came through quite quickly but ... no pistol. They had forgotten to put the weapon in the pilot's charge and the aeroplane had taken off without it.

There were frantic phone calls to London airport and a special aircraft was chartered to deliver the weapon. I explained, very vocally, that my presence on the island was valueless to the lioness in question without that particular weapon. I did offer the alternative to the customs officials – they could come and drag the lioness out for me if they wished, but I would prefer to do it my way. They agreed.

Meanwhile, I drove to the zoo and surveyed the site. It was going to have to be an open-air operation because, at that stage of the development, the zoo didn't have its own veterinary quarters, as it does now. The lion enclosure was a big open site with a central island constructed of large slabs of rock overlaying each other with tufts of turf and so forth. The slabs were so arranged that there was a lair underneath, into which the lions could go to get out of view if they wanted to be private.

Standing on the outside of the paddock, Gerald Durrell told me that the lioness was on her own in her lair and that she'd been there for some time. They were extremely worried because she hadn't come out with her cubs. The other occupants of the paddock had been separated and put somewhere else in the meantime.

I arranged for two catering-department tables to be put inside the paddock once we had got the lioness out. We also ran an electric cable with a cluster of lights suspended above it, because by this time it was obvious that the whole operation was going to run into the twilight.

The next move was the frightening one. It was necessary for someone, namely me, to go into the paddock and approach the cave in order to fire an anaesthetic dart into the lioness before we could get her out. Literally going into the lion's den. The only way I could feel at all secure would be to have a rope around my waist, and as I walked across the paddock to the mouth of the den carrying my pistol, they would pay it out. If there was any trouble, it was my fervent hope that they would pull quickly enough to get me out of the den before the lioness caught me.

I went into the paddock, feeling none too confident, and slowly walked towards the mouth of the cave. The closer I got, the louder the snarling sounded.

When I got to the entrance I lay down and peered inside, but I could see nothing except two golden head-lamps, which flashed at me from the dark distance a few yards away. Here was my lioness. I couldn't see her body at all in the total darkness. My only clues were the two golden eyes staring through the darkness at me in a threatening fashion. I lay quivering in my horizontal position, trying to aim my dart pistol at an area that I supposed would be the right place to strike the animal in the bottom of her neck.

I fired. 'Phut' went the pistol. The thud of the dart landing coincided with a hateful snarl. The moment this happened I shouted, 'Pull!' to the people outside the den.

I need not have worried, no movement occurred at all, and quite quickly the snarling stopped and total silence prevailed. I estimated it would be fifteen minutes before total unconsciousness occurred and it would be relatively safe for me to go back. After careful timing, I re-entered the paddock, complete with rope around my waist but with a following rope that I took with me in order to tie it to the lioness.

I got to the entrance to the cavern and all was silent and still. Sweating with fear, I inched my way into the darkness, pulling the rope with me. After what seemed to be a long time – it was probably only two minutes but it seemed like a fortnight – I felt a large furry paw with huge claws. It was the upper front leg of the lioness, who was now lying unconscious. I tied the rope around the foot of her leg and retreated from the den, instructing the assistants to pull the lioness out. She was towed out peacefully asleep and completely relaxed.

We quickly rigged up the tables in the paddock, covered them with sterile drapes and lifted the lioness onto them. I then clipped and shaved the whole of her flank and thoroughly disinfected it. In the meantime I had intubated the lioness and connected her to the gas apparatus. Anaesthetic gas flowed regularly and kept her unconscious. The operation site, swabbed with iodine, was isolated with green cloths, clipped at every corner, and we wore green sterile surgical gowns.

The operation began. I opened the abdomen and pulled out the uterine horns. From one horn I took out a dead cub, followed by a second one, which had gone across the exit, thus blocking it for the rest of the litter. In the other horn I was able to take out two live cubs, which began to create and call the moment they were released and put into swaddling. My task now was simple: to sew up the wounds in the uterus, muscles and skin as quickly as possible while somebody else looked after the cubs.

The operation was going well when every midge in the whole of Jersey seemed to arrive and swarm around me, the lights, the operating table and everything else. Luckily, we were towards the end of the job. We just itched.

The lioness was taken off the anaesthetic apparatus and transferred to a recovery den that had been made ready. All she had to show for the operation was an incision about eighteen inches long and about thirty or forty neatly tied sutures. The initial injections I had given her kept her deeply anaesthetized for some hours, which is what we all wanted.

By ten o'clock that night I was pretty exhausted but very

happy with the outcome. The two cubs were mewling and squeaking and drinking avidly from a bottle.

Gerald Durrell kindly took me out to dinner although I think I would rather have gone to bed.

I had to get up very early the next morning to catch the flight back to London, and at about half past five I went round to see the lioness, who was sitting up and looking very well. The cubs were thriving and the job seemed to have been completed. A car took me to the airport and I caught an early flight back. I arrived at my own home in London in time for lunch, and that afternoon I continued pruning my roses.

Gerald Durrell rang the next day and said, 'As a reward, we have decided to name the male cub after you. It's going to be called Oliver.' Now I come to think of it, quite a few zoo animals were given that name.

TAIL PIECE

16

A LEOPARD IN THE CABINET ROOM

WE HAD SEVERAL ANIMALS PRESENTED TO LONDON ZOO by the Rt Hon. Winston Churchill, the then prime minister of Great Britain. Visiting dignitaries from abroad often presented him with an animal, knowing his fondness for them, and they usually finished up in the zoo, where, incidentally, he was a fairly frequent visitor.

One day the president of a Middle Eastern airline flew in on an official visit. He was president of a wealthy company and had his own huge four-engined jet to travel in.

What the officials at Heathrow – and anybody else for that matter – didn't know was that he had brought his pet leopard with him, which was a sub-adult animal that he had reared from a cub and which went everywhere with him on a collar and chain. The huge aircraft taxied up to its appointed area, where a great crowd awaited the official guest – a crowd consisting not only of all the airline officials, but of government departments and so forth. The president of the airline was hoping to do business in Britain and thus was accorded VIP treatment.

The steps were placed in position and the aircraft doors opened. A few officials came down first and, after a pause, the president himself appeared and waved to everybody. He then descended the steps towing with him, to the astonishment of the crowd, a leopard on a collar and chain.

A customs official rushed forward and mounted the steps, barring his progress. He explained that under no circumstances could the animal touch the ground because of quarantine restrictions. A short argument occurred, but the customs official stood his ground. This caused bewilderment, clearly seen in the president's face but, since he came from a country where rabies was endemic, the customs official was absolutely right. The only alternative was for the president to return his animal to the aircraft and leave without it. This he declined to do. The impasse was solved by the president's quick thinking: 'Ah,' he said, 'I wish to present this leopard as a token of esteem to the animal-loving prime minister of your country. A gift from my country to your country.'

This altered the whole situation. The customs official sent for a crate, into which the leopard was unceremoniously bundled, and transported it to isolation quarters.

The president then continued with his official reception.

Meanwhile, the prime minister was contacted by telephone and accepted the gift with alacrity. His orders were, 'Thank the president and send the leopard to London Zoo.'

The first I knew about all this was a phone call to my office with the message that I had to get to London airport quarantine station as quickly as possible to collect a leopard which was a personal gift to the prime minister.

I collected an estate car from the zoo's transport department and set off for the airport, where I drove round to the customs sheds and the isolation area. I was greeted with enthusiasm and eagerness: they wanted to get rid of the leopard as soon as possible because they were terrified of it. The crate was equipped with little wheels, and it was a simple job to wheel the animal to the estate car, still wearing its collar with its lead attached. The young leopard appeared to be about the size of a golden retriever and was apparently quite unperturbed.

As I set off back to the zoo I was hailed by a senior customs official, who said, 'We have just received a telephone call from Number Ten Downing Street; you are to go back to the zoo via Number Ten because the prime minister wishes to see the leopard.'

Who was I to argue with an order from the prime minister? So off I went, and when I arrived I parked outside the door in front of a rather disbelieving police-man. I went round to the back of the car and opened the estate's boot and hauled out, with a little difficulty, my box on wheels containing the leopard. I advanced to the front door of Number 10, and the officer, extremely politely, said, 'Hang on, guvn'r. What you got in that box?'

'Oh,' said I, 'a leopard.'

There was one of those silent pauses while the officer stared at me and stared at the box with his hands firmly behind his back and said, 'Less of this leg-pulling, guvn'r. I repeat, what have you got in that box?'

I said, 'Officer, I have brought a leopard to see the prime minister.'

At that moment, and in response to me pressing the bell, the door opened and the prime minister's personal secretary appeared and said, 'Thank you so much for taking the trouble to bring the leopard to see the PM; he is most anxious to see it. Do come in.' So, to the astonish-ment of the London police officer, I trundled with my box on wheels into the main hall of Number 10. Staff gathered round me and the leopard, but I forbade them to touch it, not because it was unreliable – it seemed, in fact, very tame – but because the quarantine laws made it illegal. I explained that if they did touch the animal they would need a painful course of rabies vaccinations.

Mr Montague Brown announced that the prime

minister would be down in thirty minutes, because at the moment he was resting. We had a time check every five minutes from twenty-five down to five. Excitement mounted as the time drew near for the PM to appear. Exactly on cue, he emerged in his famous boiler suit, smoking a large cigar. He came down the stairs and passed through the crowd, which parted to let him come straight to me. He stared at me for what seemed a long time but was only a few seconds – he had piercing blue eyes – and growled in a friendly way, saying, 'Bring him into the Cabinet Room. Follow me.' I followed the prime minister, towing the leopard in the box, and, once inside the Cabinet Room, we shut everybody else out. He sat down and told me to sit beside him. Having settled himself into his chair, he turned round to me and, removing the cigar for a moment, said very politely, 'Please could we have the leopard out on the table.'

I explained to the prime minister that the animal was in strict quarantine and, therefore, needed to stay in the box and not be touched. After all, I thought to myself, he makes the rules and ought to know them. He turned his famous gaze upon me and growled, 'I said, please put him on the table so that I can see him.' I mentally shrugged my shoulders and thought, Well, he's the boss. I'd better do as I am told.

I opened the crate and led the leopard out on his collar

and chain. The animal seemed quite at ease and was obviously accustomed to being with people. Taking my courage in both hands, I picked him up, as you would a dog, and placed him on the table in front of Churchill. The leopard took it upon himself to stroll gently up and down and investigate the various bits of wiring that connected the microphones. I held my breath, wondering what havoc he was going to wreak. Churchill put his

hand out to stroke the animal, which responded with a sharp tap on the back of the prime minister's hand to which he grunted, 'Ouch.'

I explained that the leopard was in strange surroundings and perhaps would be best left alone as I couldn't guarantee his good behaviour for long.

'Very well,' growled Churchill. 'What I would like to do now is bring my poodle in to meet the leopard.'

I was quite stunned by this suggestion and managed to stutter, 'Sir, leopards are inclined to eat poodles, so I don't think that's a very good idea, with respect.'

'Oh, very well then,' growled the PM.

I indicated that it might be a better idea to put the leopard back in his box in case he began to get fractious or disturbed, to which Churchill agreed. I lifted the leopard off the table and put him back in the box and, being the amenable creature he was, he gave me no trouble. Once he was safely secured I breathed a sigh of relief.

The prime minister had obviously pressed a button under the table, because a man arrived with a tray of bottles and glasses. I can see the glasses now – two huge crystal buckets they seemed to be to me.

Without a murmur the servant poured the PM two fingers of brandy into his crystal glass. Churchill turned to me and asked, 'Whisky?'

I said, 'Yes, sir. A small one please.' I was conscious of the fact that it was only five thirty in the afternoon and I had yet to get the leopard back to the zoo and finish my day. Two fingers of neat Scotch were poured into my crystal bucket and it was passed to me by the prime minister. As I clutched the glass rather nervously, Churchill looked at me, raised his glass and said, 'Cheers,' draining the brandy to the bottom of the glass.

I am not used to drinking whisky so fast, or indeed at all, and certainly not at that early hour. However, I thought to myself, Drink the stuff and think of England! So I did my best to swallow the whisky in a gentlemanly fashion, which I didn't achieve because I spluttered.

My eyes began to water a bit and I felt myself going red.

Then the servant gave the prime minister another shot of brandy. The PM turned to the man and said, 'Give him another one.'

Winston Churchill turned to me with a wicked glint in his eye, just in time to see me beckoning to the servant and begging him to give me a tiny drop. The servant did his best, pouring me a small Scotch, but Winston rounded on him and said, 'That's a boy's drink. Give him a man's drink.' So he topped it back up to two fingers of Scotch. Same routine – 'Cheers. Down the hatch.' I did so with some difficulty, and within ten minutes I felt light-headed.

The prime minister was extremely kind and warm-hearted and talked about animals in general and the lion he already had in the zoo. I made my excuses, explaining that I had to get the leopard back to the zoo to install it for the night and, with his permission, I would take my leave. This he granted immediately and rose to his feet, thanking me again. He led me to the door of Number 10, holding the door open as I went outside onto the pavement with the box containing the leopard. He bade me a final farewell and shut the door.

I stared at the officer on the door, who stared back in equal silence. All I could think of to say was, 'I said I had a leopard in the box. Good night, officer.'

I drove back to the zoo and put the animal into the quarantine station.

The leopard grew to be a fine specimen and was Winston Churchill's pride and joy.

TAIL PIECE

17

BUDGIE WITH A LUMP

THE LONDON ZOO HOSPITAL HAD TO BE PREPARED AT all times to operate on any creature that needed its help, with the exception, of course, of domestic animals. We had a thriving outpatients department and never knew what would turn up. We had our own population of 6,000 animals of various species, any of which might require assistance at any moment. We had large patients – the size of elephants, giraffes, llamas – but we also had tiny patients, like rare cockroaches suffering from an infection cured with penicillin on sugar. Variety was the spice of our life and it kept us on our toes.

We had lots of small birds to treat, both in our own collection and from outside. We developed special anaesthetic techniques and constructed special instruments to go with them. Our instrument kit was made up of small tools that are normally used for operating on eyes. We called it our dolly kit.

We learned how to operate on tiny animals like mice and birds, realizing that speed was of the essence because they take fright if handled, and this is enough sometimes to kill them. We had small anaesthetic boxes made of plastic in which the animals could be placed and rendered unconscious by a flow of gas without the need to handle them; this proved very successful. After this initial anaesthetic, the process could be continued with a small tube applied to the face.

The important thing was to learn a technique for rapid surgery. We quickly realized that our whole operating time must not exceed three minutes because that is as long as small animals can withstand. We had to practise with small scalpels and scissors. In many cases we didn't insert stitches, but closed the wound by using an ordinary Kirby grip. This was both speedy and effective.

We learned that budgerigars could be operated on quite successfully providing these ground rules were carefully obeyed. The discipline of my surgical team was immense, and when we gathered round the table to do such an operation, we all paused before we started to make quite sure we were absolutely ready for what was going to happen. I had a marvellous hospital superintendent, Alec Wilson, who was not only green-fingered but extremely skilful. He was a big, strong man, but the delicacy of his touch and fingers had to be seen to be believed. I would stand with my scalpel poised and he would stand with a swab poised and I would say, 'Go!' I would then make the incision and carry out whatever operation we were performing and quickly close the wound.

I had a telephone call from my friend Dr Scales at the Royal National Orthopaedic Hospital. He had a relative – a lady living in Devonshire – who had developed a lump in the upper bone of one of her legs. Over a period of time, treatment had not worked and, sadly, the leg had to be amputated. As is the case with most amputees, she quickly regained her morale, and learned how to use her artificial leg with great aplomb. In fact, you could hardly tell she had an artificial leg, particularly when she wore trousers.

She returned to her home in Devonshire, where she lived alone with her little pet budgerigar. This bird was her sole companion, and over a period of two or three years it had learned to talk and communicate with her in a remarkable fashion. He had learned the usual phrases, such as his name and address which he would repeat over and over again. He was a very endearing little character. When the owner got back home, she was so proud of her artificial leg that she used to say to the budgerigar, by way of conversation, 'Would you like to see my operation?

Would you?' Quite quickly the budgie picked up this phrase and used it to amuse visitors.

One day the lady noticed that the budgerigar was developing a large swelling on its chest and belly. Feathering had covered it up for a while, until suddenly it became obtrusive. This alarmed the lady; she felt quite sure that she had given her cancer to the bird, which, of course, was totally untrue, but nonetheless she was shattered by it, and rang her relatives, the Scales family, to tearfully explain the problem.

My friend then rang me, and I arranged to have the bird brought up from Devon and admitted to the zoo's hospital. When I examined the budgie, I found a large fatty tumour attached to his abdominal wall. Operation was the only hope.

My friend said, 'Surely you can't operate on a tiny thing like that?' He, of course, was used to operating on large human limbs and putting in steel bits and pieces.

'Oh yes we can. We've practised for a long time and have become quite good at it.'

The bird was left with us, and we anaesthetized it in one of our small boxes and stripped the feathers off its chest and abdomen to prepare it for operation.

The operation proceeded very quickly and was totally successful, leaving a little row of sutures down the chest, which we decided, on this occasion, because of the size of what had been removed, would be the safest thing to do.

I rang the anxious relatives and told them that the operation had been successful and that the fatty tumour had been removed and the budgie was sitting up on his perch again. They were amazed and relieved. I did warn them, however, that he would need to stay in hospital for two or three weeks until we could remove the sutures, but that there was no longer any danger of infection.

At the appointed time, three weeks later, my friend called to collect the bird for his relative. I decided to tease him.

He said, 'I still can't believe that you operate on these tiny creatures; it seems like magic to me.'

I couldn't resist the temptation, so I said, 'You don't really believe we operate on them, do you?'

He was astonished and said, 'Well, how on earth have we got the bird back?'

'Quite simple,' I said. 'We have a room out the back of the hospital containing a thousand budgerigars of different hues and colours. When we are faced with this sort of problem, we wait three weeks and issue a replacement bird once we have taught it to say all the phrases the original bird said.'

He swallowed this. He believed it so sincerely that I couldn't change his mind.

The story had a happy ending, though. The bird went back home to Devon, to the joy and delight of its owner, and it already knew the phrase, 'Would you like to see my operation? Would you? Would you?' Which now, of course, it could apply to itself!

The surgeon who had operated on the lady paid her a visit while down in Devon to see how she was getting on with her artificial leg. He was ushered into the front room where the cage containing the little budgerigar was located.

He had no idea the bird was there as he gazed out on the Devonshire countryside, until suddenly a little voice said, 'Would you like to see my operation? Would you? Would you?' He thought the place was haunted and ran out of the room to find his patient and said, 'I've just heard a ghost!'

18

A PELICAN IN THE
WAR OFFICE

THOSE WHO KNOW THE PONDS IN ST JAMES'S PARK MAY
have seen the so-called 'royal' pelicans on display there.
The monarchy has always had a soft spot for pelicans,
and they rather like to see them floating about majestically with their extraordinarily voluminous bag-shaped
lower bills. One of the most dramatic sights is when
the birds come out of the water, waddle up onto a rock
and stretch out their enormous wings in the sunshine.
This sunning process not only freshens up their feathers
but plays an important part in the metabolic process of
converting vitamins in their bodies. It looks most
attractive and is a grand sight when several display
simultaneously.

There is a problem, however. The pelican's enormous
wingspan gives it tremendous lift for flying and if they
just open their wings in a strong gale they can lift off the
ground and fly quite easily. Within the confines of St

James's Park, this is not desirable, and for many years the zoo was called upon to pinion a wing on each of the birds.

The Royal Household expressed some concern about this, and were upset, I am told, at seeing one of the wings short of part of its anatomy. This was particularly visible, of course, when their wings were stretched out for sunning. A message was passed down to see whether or not an alternative system could be thought of.

There is an alternative system, but it isn't practised a great deal because it's so much simpler to pinion them when they're chicks. The alternative consists of removing one or two strips of the extensor tendons on the leading edge of the wing – these are similar to the tendons that operate your fingers if you look at the back of your hand and move your fingers up and down. The idea is to remove a strip of the tendon, which interferes with the bird's ability to thrust downwards into the wind for flight. This, in theory, makes them captive, but doesn't show as a mutilation. It was agreed that this should be done, and under anaesthetic a tendonectomy was performed on several of the birds. They rapidly healed and were able to go back on exhibition in St James's Park.

This pleased everyone because, when the birds sunned and their wings were fully extended, there was no sign of any mutilation or abnormal anatomy. We were all quite chuffed and quietly patted ourselves on the back for having achieved this one-off job.

Our joy was comparatively short-lived, however. A mini whirlwind of the sort created among the high buildings in London by sudden gusts of wind struck St James's Park. It whisked up all the leaves and waste-paper baskets, and flung everything all over the place. Two pelicans took off in the wind and floated majestically upwards, circling the tops of the trees. The first one ran out of steam and did a magnificent crash landing in Birdcage Walk, much to the astonishment of passing traffic. It was quite unharmed and waddled back to the pool from which it had come.

The other bird went a bit higher and got caught up in a vicious side draught, which took it flying up, way over Horseguards Parade. Totally out of control, it flapped and headed on a down draught for the nearest building, which happened to be the War Office. It crashed through some large windows, interrupting a meeting of the foreign Chiefs of Staff. Imagine the shock of a huge pelican diving in through the windows and crash-landing on the table in front of these astonished gentlemen. There was a fearful scuffling and cries of alarm in various languages.

The zoo was telephoned and asked to remove the bird immediately and check what had gone wrong with the operation designed to stop it flying.

The truth of the matter was that the force of the wind had been exceptional that day, but we hadn't realized that the severed ends of the tendons could, and did, fix onto the bones and exert a similar strength to the original tendon.

I have no idea whether the incident altered international policy at all, but anti-aircraft defences were strengthened afterwards. The aerobatic pelican survived his crash-landing and was returned to the island in St James's Park, where he continued to sun his wings.

THE IN-HOUSE LION

THE MAIN OFFICES OF LONDON ZOO ARE IN A MOST imposing Georgian-style building with a pillared front door, and you approach it by a semi-circular drive from the Outer Circle in Regent's Park. As you enter the imposing hall, the famous meeting chamber of the society is straight ahead of you. The entrance is guarded by two huge mounted tusks, which were collected some hundred years ago, before the collection of ivory was forbidden. At the reception desk there were always two or three people on duty, including Sarge, a six foot six ex-guardsman who intimidated everyone by his sheer presence.

My office was on the right-hand side as you came in, as far away from the front door as possible. There were two doors leading to my office, one to my secretary and the other to the superintendent's office, and through the windows over my shoulders I had a view of the canal banks.

One day, through the main doors and into this paradise

of dignity and silence, guarded by Sarge at the reception desk, strode a large male lion – a magnificent beast – attached by a cord to a man who walked up to Sarge at the enquiry desk and asked to see the vet. An emerging secretary scuttled back into her office, and in seconds the whole building discovered, with baited breath, that there was a lion walking around the main hall. Slowly, Sarge reached for the internal phone and dialled 19 – my number. His singularly powerful stentorian military voice sounded very high-pitched and squeaky as he said, 'Sir, there's a lion in the hall to see you.'

A hush had fallen over the whole building; no-one moved or apparently breathed. Trying to appear calm, I left my office, went down the corridor into the hall and, as I turned, I faced a huge male lion complete with flowing mane. I stood very, very still. The handler called out, 'He's tame, sir. He won't do anything.'

Reassured, I went up to the man and said, in a quavering and somewhat nervous voice, 'What the hell do you mean by bringing a lion in here like this?'

'He's ill, sir. He won't work.'

'Where does he work?'

'Elstree, sir. He's in a film, and because he won't work they've stopped shooting, which is costing thousands of pounds a day, so they told me to bring him here.'

I was amazed and asked him how he'd got here.

'Simple,' said the man. 'I just put him in the back of my Bedford van and drove him down here to see you.'

By this time, even Sarge was standing rigidly to attention and hardly daring to breathe; my heart was doing about 180 beats a minute and the palms of my hands were sweating. I told the man brusquely to put the lion back in his van and I would take him down to the animal hospital.

So it was that I found myself sitting in the passenger seat of an old Bedford van, with a lion handler driving and a huge tame lion breathing down my neck.

Once at the hospital, in a secure, large chain-link cage,

I was reassured again by the handler that he was as gentle as a lamb. I asked the man how he knew he was ill.

'Simple,' he said. 'He won't pounce.'

'He won't pounce?' I said.

'Yes, sir. He won't pounce. He just lies down and grumbles and growls and whines. He certainly won't eat, and he goes on moaning.'

I realized I needed to examine this beast, and was reassured again by the handler that he was extremely tame and had been trained like a dog. I asked him to get the lion to lie down; he did so and the lion quietly obliged. On my hands and knees, I palpated the lion's vast abdomen, saying lots of silent prayers. I identified a painful spot, which produced a grumble when I pressed it. It was in the region of the lower bowels. I asked the handler to stand the animal up and lie him down on the other side; he did this and the lion obliged. Astonishing! I palpated the other side and found exactly the same problem; the lion obviously had an intestinal blockage that needed shifting.

I asked the handler to stand the animal up and hold on to his head, while I walked down the other end and took the lion's temperature. I thought he might be developing a virus or infection. The routine way to find out if an animal has a fever is to insert a thermometer in its bottom. So, grasping the root of the lion's tail in my left hand, I gently inserted the thermometer. He didn't like this at all. In fact, he took off, towing the handler with him, together with myself, hanging on to his tail, anxious not to lose sight of my precious thermometer. We did a couple of laps of the huge cage in this fashion, and the lion didn't stop until I'd removed the thermometer. He was, I guess, aggrieved and humiliated. His temperature was 105 degrees Fahrenheit. Not much of a fever, if at all, because the exercise of rushing around the cage probably pushed it up a bit.

I told the handler that the lion was constipated, and probably had a lot of material blocked up, which needed to be got rid of quickly so that he would feel better and return to work. I gave the lion a powerful purgative injection which, I warned the handler, might take effect before he got back to Elstree. I also gave him some laxative medicine to put in his milk, which he claimed he could simply pour down the lion's throat. I believed anything by this time! The handler got back in the Bedford van with the lion and shot back to Elstree. He promised to let me know how he got on.

I didn't get a telephone call the next day, but I did get one forty-eight hours later. The man said, 'Do you want the good or the bad news?'

I said, 'For goodness sake, give me the good news first.'

'Well,' he said, 'he's a lot better. He's back on his feed, and today I am happy to tell you he pounced.'

'Pounced on what?' I replied.

'Oh, we have a sort of dummy man that he pounces on in the film. Now he has pounced again the filming can continue.'

I said, 'So much for the good news, what's the bad news?'

'The bad news is that the injection you gave him worked halfway to Elstree and he emptied his bowels all over the van, all over himself and partly over me and the seats in the cab. Very effective stuff, that injection.' He sounded a little bitter. 'Do you realize it took me the whole of yesterday to clean up the van and the lion ready for filming? All the technicians threatened to walk out if the animal didn't smell a bit sweeter.'

The in-house lion got better and filming commenced, but I never got any free tickets!

20

GORILLA WITH A HERNIA

IN THE EARLY DAYS OF THE NEW ANIMAL HOSPITAL AT London Zoo there were many rules and regulations which needed to be set up and obeyed. There were essential rules about security for the safety of the staff as we never knew what was coming in, be it a tiger or a mouse. I had designed steel sliding doors, operated by steel handles which slid up and down and pulled on steel cables. The orders for the day called for the handles to be padlocked into the particular open or shut position so that no animal could alter the position by tugging on the door. These sorts of precautions were essential.

Apart from the rules and regulations over health, hygiene and security, there were other rules, too. For example, all staff during working hours were to wear full protective clothing and face masks in the hospital corridors or other places where animals might be. Some people queried this as unnecessary, but it was far from unnecessary. Nobody had any idea then, or even now, what curious diseases these animals from foreign countries might have or be excreting or coughing. In order to protect my staff, I insisted that they wore protective clothing and face masks. Other research units that handled monkeys accepted my advice and even equipped their staff with 'spaceship' helmets so there could be no danger of the staff getting drops of infection from coughing monkeys. I make no apology for being over-particular.

Another rule applied to access to the hospital from outside. As soon as it became known that the hospital existed, there was a flood of enquiries from a wide variety of people for us to see their sick animals.

At the time we had no outpatients facilities, and I was instructed by Solly Zuckerman that under no circumstances could animals from outside be admitted to the hospital for any purpose, and certainly not for surgical procedures. This was a savage restriction and one I didn't agree with but had to obey. I received written instructions to support this and was therefore surprised when one day I got a letter from a lady who looked after young gorillas for a famous zoo, asking me to examine and, if necessary, operate on a young gorilla that had a hernia in the middle of its abdomen. I advised the lady that I was quite unable to assist because the then secretary of the society had forbidden me to see outside animals. A few days after I had despatched the letter, I got a violent response from the secretary, who ordered me to examine the gorilla and, if necessary, operate on it. I did this willingly, but was curious to know why the situation had changed so rapidly.

I duly made an appointment for the lady to come to the hospital. Imagine my surprise when an enormous Rolls Royce drew up outside the hospital door, driven apparently by nobody at all. The chauffeur, who turned out to be a very small man who sat on a cushion in the driving seat and drove the vehicle by peering through the steering wheel, hopped out. When he got out, he rushed round to the side and opened the door. To our astonishment, the lady got out of the car carrying a sub-adult gorilla in her arms. She reassured me that the gorilla was friendly, and that, yes, she looked after several of them and they all had their meals together. I nervously enquired where she kept them in her house and, in particular, where they slept?

'No problem,' said the lady. 'They sleep in the bed with me.' I well remember how, at this point, my then secretary, Miss Mary Stanford, a charming young woman, blushed to the roots and fled, notebook in hand. Left as I was with the hospital superintendent, the lady and the gorilla, I proceeded to take some notes.

The gorilla was admitted to the hospital for further examination and a variety of tests before the operation. It was important for me to establish that the health of the animal was good enough for him to withstand a surgical procedure. This meant days of throat swabbing, blood, faecal and urine testing and radiography.

In the interim, being a little nervous of operating on something as human as a gorilla, I rang a surgeon colleague of mine, Mr Atwill, at Great Ormond Street Hospital for Children. He was, as all my medical friends are, extremely co-operative and helpful, and he offered to examine the patient. He thought the operation posed no particular problem, providing I was able to administer a suitable anaesthetic. He would assist me on one side of the table, while I operated on the other.

On the appointed day he arrived in good time. By this time the gorilla had received premedication and was quiet and sleepy. I pushed a tube into his trachea and down into his lungs and attached it to the anaesthetic gas machine to keep him thoroughly anaesthetized. We then cleaned and prepared the operation site, scrubbed up and gowned up and came back into the theatre to start operating. It was a joy to see the skill of my colleague. He said, 'It's exactly the same as operating on a child; there are no differences at all except there's more hair.'

The operation proceeded successfully and the animal

finished up with a row of sutures across the top of his belly button. He was, at this stage, still unconscious. We mused on how on earth to stop the gorilla interfering with his stitches. I knew that if we tried to bandage the wound the gorilla would rip it off. If we put sticky tape over, he would spend hours tearing it off, together with the sutures. And so the conversation went on.

In the end, I decided to match our intelligence with that of the gorilla. I put a plaster-of-Paris cast on his left foot and halfway down his leg, then I bound this with copper wire and put more plaster of Paris on and wrapped the whole lot in sticky elastic bandage so he had an enormous club foot.

It was amusing to watch the gorilla spend four days desperately trying to get the stuff off his foot and paying absolutely no attention to his abdominal wound. By the time he realized he had been tricked and had got all the stuff off his foot, the wound in his abdomen had healed and we could take out the stitches. The gorilla made a straightforward recovery, went back to his home and grew up to adulthood, fathering some baby gorillas.

In the meantime, common sense prevailed and I persuaded the society that we needed to have an outpatients' department to deal with this and other similar matters that might arise. It became a very popular feature of hospital life.

21

TRANQUILLIZING FIFI

FIFI, A HUGE FEMALE HIPPOPOTAMUS, USED TO LIVE IN the hippopotamus pond in London Zoo. She was quite happy there and wallowed about in what she considered to be her delectable mud. Because my flat in the zoo was situated between the hippopotamus and the seals, I was not always happy about her mud when the wind blew in my direction. It was decided that Fifi should be transferred to Whipsnade Park, where she could join a male hippopotamus called Neville, and where everybody hoped she would subsequently produce some young – always a great draw to the public. Fifi's journey to Whipsnade was quite uneventful, and in due course she found herself in a large pool in Whipsnade Park, next door to the one occupied by her prospective mate, Neville.

After a period of acclimatization and getting to know her neighbour, the two animals were introduced. There was a lot of sloshing and sploshing for quite a time, and a fair amount of roaring and squealing and shouting, but they seemed to settle down together quite amicably. From time to time they were seen courting heavily and everyone hoped that Fifi would become pregnant.

However, the normally fairly placid Fifi became very neurotic, somewhat hysterical and violently aggressive. The keepers noticed this because, instead of receiving them in a friendly way when they took food in, she charged and roared and made a frightful fuss. Her whole change of temperament and subsequent rejection of Neville suggested she was pregnant. In fact, as events proved, she was.

Pregnant women can sometimes show untypical behaviour, and curious desires for odd foods. Fifi, on the other hand, simply became aggressive and troublesome. It was decided that Neville ought to be removed; Fifi was becoming so aggressive towards him that we were worried she might injure him. By a quirky change in feminine nature, Fifi then became much more aggressive and practically wrecked her accommodation in distress at Neville's removal. Hobson's choice! Neville was returned to her in order to calm her down, but, on the contrary, she attacked Neville so violently that he was injured and needed to be separated again. We all realized that something had to be done to quieten

Fifi down in preparation for the imminent birth.

I could find no literature anywhere to tell a lonely veterinary officer how much sedative to give a hippopotamus, or how to administer it. We guessed that Fifi weighed about two tons, perhaps a little more, so we gave her a daily dose of a drug called promazine hydrochloride – fifty times the quantity of a human dose. We mixed the drug with Fifi's evening meal and the effect was magical; she calmed down and became friendly.

However, we didn't risk putting Neville back in with her since we thought she was too close to having her baby.

It is difficult to know when a huge two-ton hippopotamus cow is going into labour, and equally difficult to know when she is actually giving birth. In the end, she had her baby overnight under the eyes of an infra-red camera, and was an extremely good mother. Unfortunately, once her new baby arrived she became violent and aggressive again, and tried to attack the keepers and her stalls. She was again given the sedative, and the baby-rearing progressed without much more trouble. For the next six weeks she and the baby were kept indoors, going in and out of the indoor pool, and the baby fed from her mother frequently.

Eventually we removed all sedatives from her food and prepared to let the mother and baby into the outside lake so they could swim about and the public could see them. On the appointed day, early in the morning, the gates were opened and, much to everybody's surprise, Fifi took her baby down to the water's edge and quietly slid in, proudly showing it off. Neville, the father, was near by, and he swam over to join them, taking great interest in the baby. Fifi was extremely careful not to let her husband get too close to her precious baby or interfere in any way, but otherwise they made a very happy family scene.

*It seemed that married life and the presence of
a baby were a sufficient tranquillizer for Fifi.
She needed no more medicine in her food.*

THE CORNISH MOUSE

DURING THE COURSE OF MY ADVENTUROUS LIFE AS senior veterinary officer to London Zoo I gained unique and wide experience of the veterinary problems of the non-domestic animal. My patients ranged from a rare cockroach that had a throat infection, which we cured with penicillin mixed with sugar, to huge elephants and other large mammals. The zoo contained all forms of life – birds, fish, insects, reptiles and four-legged animals of all shapes and sizes – and they suffered from myriad problems that needed to be treated individually.

It was small wonder, then, that some veterinary surgeons in general practice, when presented with a curious creature not in their normal textbooks, like snakes, pottos, bushbabies or marmosets, which were rapidly increasing in numbers as pets, contacted London Zoo for advice. The increase was probably due to the popularity of the early TV zoo programmes given by the then superintendent, George Cansdale.

The demands from the enquiry section became so great that they were taking perhaps too much time out of the working day. The outpatients' clinic, situated in the new zoo hospital, solved the problem. There, veterinary surgeons could refer to me cases that they themselves felt unable to grapple with. The system worked very well and was frequently in use. We rarely had less than four or five referred cases first thing every morning. They were wide-ranging; on one occasion we were delighted to welcome a professional snake charmer, who had kept her snakes in a big wicker hamper at the Palladium until the electric heating pad in the hamper had burned the snakes. They made a complete recovery in the hospital, after which they returned to her snakecharming act. Lots of children brought in their pets; they were usually in a state of great anxiety because they were fearful of what might happen if the animal remained untreated.

So it was that one day, as I was walking across the main hall of the new hospital to go into the administration section of the operating theatre, I noticed a father and daughter sitting patiently in the waiting area. She was perhaps seven or eight years old, and on her lap was a small cardboard box, which she held tenderly between her hands, and gazed at constantly. The little girl's legs dangled from her chair, not quite reaching the floor. She was a sweet sight, with neatly brushed hair held back in an alice band, a white blouse, dark-blue skirt and patent-leather button-across shoes.

I returned to my office and buzzed my secretary to find out the details of this next case. A mouse!

The father came in with his tremulous daughter and they both sat in front of me to talk about the mouse.

The father said, 'This mouse is my daughter's precious pet. He's called Mickey, and he seems to have developed something rather unpleasant underneath. We want to know if you can cure it.'

The little girl looked up at me anxiously, and I said to her, 'Tell me, what do you think is the matter with your mouse?'

'Well,' she said, 'it has grown this horrid lump and it looks very nasty. Please will you cure it.'

We put the box on the table and I carefully opened it so that I could grasp the mouse gently in my hand, with its head between my left thumb and forefinger and my fingers cradling the rest of its body upside down so that I could get a view of the abdomen. I couldn't show it, but I was appalled to see that a very large breast cancer had grown to be almost half the size of the tiny mouse.

I had to decide how to tell the little girl the mouse

needed to have an operation and a general anaesthetic, and that, because it was so tiny and frail, it was unlikely to survive surgery. When a cat pounces on a mouse it is often dead before the cat strikes, from a heart attack from sheer terror.

However, we had gained a lot of experience over the years using our dolly kit of surgical instruments and a special miniature anaesthetic apparatus. These instruments were really those used for eye surgery and were therefore very precise and small. This had given us an extraordinary success rate at removing tumours and mending broken legs in budgerigars, and other such delicate tasks.

I turned to the little girl. Her eyes were limpid and terrified as I explained what we had to do.

She listened steadfastly and said, 'Will he die?'

I replied, 'We can only do our very best for Mickey. We will take the greatest care of him, and there is quite a chance that he will survive to go home with you.'

Tears poured down her face as I turned to her father and said, 'I'll take Mickey into the operating theatre immediately so that we can do this as soon as possible, but you do understand that there are risks?'

'Yes, of course,' said the father, putting his arm around his daughter's shoulder.

I took the box containing the little mouse out to the

preoperative room, where we prepared for an operation in the ordinary way. My assistants cleaned the
mouse's surgical area for me, and then, using a special
little gas box, we rendered him unconscious, transferring
him to the operating table, where we continued his
anaesthetic from a tube, covering the rest of his body
with a drape.

Experience had taught us that when you operate on
such tiny creatures speed is of the essence; your surgery
must be quick and accurate, and there must be no blood
loss to speak of. In fact, when operating on budgerigars
we reckoned we only had leeway to lose 3cc of blood.
Such tiny quantities are quite frightening when you're
operating.

I commenced the operation, quickly opened the skin
and subsequently was able to remove a nasty tumour,
freeing it from all the surrounding tissue. That left us
only a minute or so to stitch up the wound tightly, clean
the animal and put it into some clean tissues in its box to
regain consciousness. This it did in about three minutes
and, to our delight, it ran around its box apparently quite
happily, if a little drunk!

After a short time, I took the box containing the
mouse, which was now fully recovered, and walked
across the hall towards the anxious little girl. I called out,
'It's all right, it's all right. Mickey has done well; he's had

his operation and is sitting in his box with a row of stitches in his tummy.'

I gave her the box and she tremulously opened the lid; to everybody's delight, Mickey shot out, ran up her arm, sat on her shoulders and busily proceeded to wash his whiskers and have a general tidy up.

I discovered later that father and daughter had come all the way from Cornwall to have the mouse treated. As I operated on Mickey, I couldn't help feeling the heavy weight of responsibility. It was as if the little girl were watching my every move. I can safely say that the look of relief on her face when she saw Mickey alive and well again made this one of the most emotionally rewarding operations I have ever performed.

23

TIGER'S EYE

OF ALL THE BIG CATS, THE TIGER IS THE MOST DRAMATIC. It has vivid colouring patterns and is full of grace and beauty. I once hand-reared a cub to two years of age, and she remained as friendly as any domestic dog. I was sad when she had to go on exhibition and the bond with me was disturbed, although she still greeted me from way over the other side of the zoo if she saw or heard me. Of all the big cats, the tiger remains basically monogamous and strides with its mate through the jungle as a pair.

The zoo had a pair of large Malaysian tigers – the male was called Kahn and the female Nepti. They made an attractive pair and were very loving. Kahn was considerably older than Nepti, who had been my pet for the first two years of her life. Now she was a grown-up lady and had her own husband.

Kahn and Nepti had already produced twins – known as the terrible twins because they were always up to mischief. Nepti was a bit skittish, and on this particular morning, when the meat was thrown into the den, Kahn took it into a corner. Nepti observed this for some time and thought, Hell, why should I do without my breakfast? She flew across the den, snatched the meat in her powerful mouth and rushed back to her own corner, where she lay down, staring at Kahn, defying him.

Kahn, with the greatest of dignity, studied the matter for a moment, laboriously raised his huge bulk to his feet and strode purposefully towards Nepti. She twittered with greeting but failed to realize that Kahn was far more serious about the matter than that. As he got close to Nepti, he swiped her head with his massive paw, and as she lowered her head he stole the meat back and rushed into his corner with it. Nepti had closed the left eye that received the force of the blow.

This matter had been witnessed by the keeper who had put the meat in, and when he saw what had happened, he immediately telephoned the hospital and summoned me. I agreed with him that, at the moment, observation was the only thing to do, since it might well just be a temporary black eye.

A week or two elapsed, and my daily examination showed that the eyeball itself was swelling rather rapidly and that it had a staring look about it. We took Nepti into the hospital and, by using a restraint cage, I was able to put an ophthalmoscope to her eye, which revealed that the lens inside had disappeared out of its holding capsule and fallen into the bottom of the eyeball. This blocked drainage canals, and the eyeball was becoming glaucomatous. If the condition remained untreated it could endanger the sight of the eye, which was the last thing we wanted.

I telephoned a friend and colleague of mine, Sir Benjamin Rycroft, who was the leading ophthalmologist in Britain; he had always been extremely helpful to me regarding difficult eye cases. He agreed to come straight away, while we had the tigress in the restraint cage, to check the eye out. He arrived about twenty minutes later and confirmed that the tigress had got early glaucoma and had lost the lens out of the capsule. In answer to my question, he said, 'Yes, she will need to have that lens removed from her eye and would you like me to do it?'

Of course I wanted him to do it. Operating on a tiger's eyes was no particular forte of mine compared to the best ophthalmologist in Britain. I had to administer certain drugs to the tiger's eye for the next day or two to prepare her for operation, and our ophthalmologist said that, on the day in question, he would arrive with his team at eleven o'clock in the morning. His team would consist of himself, his anaesthetist, his senior registrar, his theatre sister and a theatre nurse. My duty was to prepare the tigress for the operation and show his anaesthetist how I administered anaesthetics to zoo animals in general and the tigress in particular.

On the day, we premedicated the tigress with an injection to quieten her down and render her tractable. We then transferred her to the operating table for further preparation, which included putting a drip into a vein in her hind leg and introducing a large tube down her windpipe to feed more gas anaesthetics into her lungs.

By 11 a.m. we had the tigress properly anaesthetized, intubated and entirely covered with green cloths, so the only thing clearly visible was a great big shiny eye peering up out of an orifice in the sterile green surgical drapes. It was a dramatic sight; the tigress weighed 300 pounds and seemed to stretch for ever; her legs overlapped the ends of the long table.

I went out to greet the ophthalmologist and his team, and being a most generous-natured man, he said, 'Oliver, this is your operating theatre, you are in charge. We will do exactly what you tell us.' We took them through to the pre-operation area, where they gowned up and put on their masks and gloves, then we entered the operating theatre. The sight of this huge sea of green, underneath which lay a 300-pound live tigress under anaesthetic, was so dramatic that it stopped all of them in their tracks for a moment while they took it all in.

I positioned myself at the anaesthetic apparatus, maintaining a suitable depth of anaesthesia for the ophthalmologist. The ophthalmologist's anaesthetist came to stand beside me and saw how much gas I was administering to the tigress. He gave a startled look and said, 'You're giving her far too much. You must back down immediately otherwise you'll give her oxygen poisoning, apart from anything else.'

I took no notice, and he repeated his demand that I turn down the gas flows. Just as wild animals need greater amounts of oxygen during an operation, so they need larger quantities of anaesthetic, as I discovered when I operated on Sabre the puma. I explained to him that such a large animal burns up gases at a tremendous pace related to their escape mechanisms, but he wasn't satisfied. In order to prove the point, I lowered the flow on one of the critical taps. Within thirty seconds one of the huge paws peeping out from under the sea of green moved slightly, and the claws on it began to move.

The anaesthetist shouted, 'Turn it back up, turn it back up!' He was obviously terrified to see that the variation might produce a live tigress bounding around the operating theatre. This drama over, I indicated to

the ophthalmologist that the animal was ready to be operated on.

Silence fell as he carefully prepared to enter the eye and remove the lens. The tension in the operating theatre was terrific. Not only was the theatre warm, but the people who normally operated on human eyes were discomforted by the sight of the huge tiger's eye and, in particular, by what lay underneath the green drapes.

The first man to faint was the senior registrar, who slumped down with a thump. Nobody said anything, but two members of my staff walked quietly over and took him outside and propped him up against a wall in the fresh air. He never returned. As the operation proceeded there was another thump as the theatre nurse hit the floor. She, too, was taken outside to join the registrar. No-one else noticed or was disturbed, and the operation progressed smoothly until a slithery, silvery lens was removed from the eye. That done, some corneal sutures were placed in position and the operation was over. The tigress made an uninterrupted recovery. Her vision was probably a little blurred in that eye, but at least she had sight. She never stole her husband's breakfast again.

Some weeks after this episode, I was standing at the club bar of the Royal Society of Medicine, having a drink with one or two friends. My ophthalmologist friend entered the dining hall, followed by a crowd of acolytes who had all been listening to him

24

FIGHTING SNAKES

EVERYBODY HAS HEARD OF BOXERS FIGHTING, OF BULL-fights, regrettably of dog fights, and of cock fights. In many countries fighting between animals for sport is banned. Many animals in the wild fight, but for a specific and real reason. There may be territorial claims being repulsed by dominant males, or skirmishes among enthusiastic young males to gain domination of some females, and so on. But rarely, very rarely, do animals take their fighting to the death. Inevitably the fight stops when one gives in and retreats, leaving the victor in charge.

Not many people will have witnessed a snake fight. They have no limbs, after all, so how on earth do they fight? A snake's locomotion is a slithering action, and the whole snake is propelled by the movement of the bands that connect the ends of the ribs.

Among its large reptile collection, London Zoo had a

very fine Indian reticulated python, a lovely snake with variegated patches and black and yellow colouring. A most attractive sight, even if you don't like snakes. Another python was introduced into the same enclosure as it was felt that there would be no harm done and it would make for a better display.

The enclosure in the reptile house contained a large pond in which the snakes could bathe, as many snakes are good swimmers. One day the keepers saw these two pythons squaring up to each other, and they didn't quite understand what was going on until, suddenly, both snakes reared up and opened their mouths wide. They flew at each other, entwining their bodies and trying to swallow each other. London Zoo has, in its museum, a specimen of a snake that had been completely swallowed lying inside another snake. Both snakes died.

Our two combatants violently started to try and swallow the other. All that happened was that the smaller snake kept grabbing bits of the bigger snake's body and, once it had latched on, it snatched its head backwards, tearing the tissue. This went on for quite some time, until the larger snake was quite badly damaged all the way down to its tail, at which point it sought refuge in the pond and the smaller snake wrapped itself around one of the tree stumps to lick the wounds it had suffered.

We brought the larger snake into the hospital for

repair work; it was a long and tedious job, but comparatively easy, because wherever the tears were we could sew up the tissue so the pattern matched. Rather like repairing a Fair Isle pullover.

The snake recovered well for the next two or three months, but then I was summoned back to the reptile house to see it; and found that it had swelled rapidly, like a pumped-up motor-car inner tube. The animal was in great discomfort and something clearly needed to be done. We examined the snake and found that its exit hole at the bottom – the cloaca – which gets rid of all the solid and the liquid material the snake discharges, had completely seized up after being damaged in the attack it had suffered from the other snake. There was only one thing for it: into the hospital to see what we could do to help.

It became quite obvious that the original hole nature had provided the snake with was completely blocked up and non-functional, hence the enormous swelling. Food was fermenting in its guts, forming gases. Instead of being about three inches in diameter, the snake was approaching five, and in some places six, inches. Something had to be done, and quickly.

We prepared to operate. In a human, the operation of opening up the guts and making an exit tube in the abdominal wall is known as a colostomy, so it appeared that we were about to do the first and only colostomy on a snake. Not surprisingly, there are no textbooks written that include instructions on how to do this operation.

A keeper held the snake every two or three feet. The area where I was operating rested on the surgical table and was prepared. The next problem was where to cut the snake. The bands that go round the snake and connect its ribs are used for locomotion, so should I cut across, like slicing a cucumber? Or lengthways, like making a banana split, severing some of the important bands? I decided to adopt the banana-split technique.

We operated and opened up the abdominal cavity. I'll spare you the details, but we removed two buckets full of evil-smelling material. My hospital superintendent told me subsequently that he had never had a better crop of tomatoes!

Having created a new hole in the wall of the intestines, we stitched the intestinal wall to the abdominal wall, giving the animal a completely new excretory orifice.

The snake was returned to its den and, to my astonishment, it went straight into the pool and swam around, washing out its abdominal contents in a very satisfactory way.

The python lived happily for a long time and eventually died a normal death, but very few of my medical colleagues believe that I performed a colostomy on a snake.

25

WHATEVER ONE CAN DO, TOUCAN DO BETTER

THE TOUCAN IS A QUITE EXTRAORDINARY BIRD, AND some believe it to be dramatically good looking. It has an average-sized body, about the size of a large pigeon, or perhaps a parrot, and its head is adorned with a most enormous bill which is coloured bright yellow, except for the tips of the upper and lower bills, which are black. It has a cunning way of eating: it takes hold of its food in the most delicate fashion with its long bills and throws its head upwards slightly, opening its mouth and letting the food travel down a specially grooved long tongue that directs it the right way. It then swallows. Thus it is important to the toucan that its bills remain in good shape and meet accurately at the end. If they didn't, the bird wouldn't be able to feed and would starve to death.

The zoo had a pair of toucans that had lived together quite amicably in the same cage for quite a long time. It was a nice cage, decorated with branches and leaves and situated in a warm birdhouse; they seemed idyllically happy, hopping about along the branches.

One day, it seems, the male made what the female considered to be improper advances to her, and she lost her temper and attacked him. With a mighty swipe, she bit off about one or two inches of his lower bill. I'm sure there is a message here for all courting couples! So then we had a handsome male toucan whose upper bill overhung his shortened lower bill by a few inches and, horror of horrors, he could not feed because he couldn't pick up his food any more.

The bird was put into a hospital cage and I decided to seek the advice of friends of mine at the Eastman Dental Clinic, which had a splendid prosthetic department. I rang them up and explained the problem. They were overjoyed to be presented with something a little out of the ordinary which would be interesting to do. We arrived at the Eastman with the toucan in a carrying cage, looking very dejected, obviously missing his bit of bill but, on top of that, probably feeling thoroughly ashamed at having been thrashed by his girlfriend.

The people at the Eastman were magnificent. In no time at all the retrieved piece of bitten-off bill was colour-matched with the remainder of the lower bill, and a suitable cast was made in dental materials to mould a plastic replacement. The next problem was how to attach

it? They made a very hard metal titanium alloy frame which would be glued onto the existing lower bill to hold the newly made plastic bill. The glue used was a dental glue acceptable to tissues and was very fast-acting.

We returned after a few days for a trial fitting of the metal frame and prosthetic bill, and, joy of joys, the match was perfect. The artistry of the colouring of the prosthetic piece was such that it was indistinguishable from the real bill, right down to the black colouring towards the end. It was amazing, and when the whole thing was fitted, the toucan not only looked perfect, but was able to pick up his food and feed properly.

What we learned after all this, apart from the skill of the prosthetic department, was that although the lower bill is mounted on a bony core – and in theory bone never regrows – the constant tapping on the end of the new bill caused the bone to grow and thrust the prosthesis outwards. So then we had a toucan whose artificial lower bill had to be continuously filed back to accommodate that growth, just as we file our nails! Eventually we were able to remove the frame and prosthetic device and allow the lower bill to function normally, which was a dramatic outcome when we had been repeatedly told by expert anatomists that the bone would never grow back and the prosthesis would need to remain there for ever.

The toucan convalesced in a cage side by side with the mate that had bitten off his bill. They seemed to get on well together and the decision was made that they could once again cohabit. After they had been together for some days, the male quite sud-denly gave the female a hell of a ticking off and knocked her around the cage for about five minutes. We were all convinced that he was getting his own back and teaching her a lesson.

26

WHO'S COCKY NOW?

LONDON ZOO IS MECCA FOR CHILDREN AND THEIR
parents. In the school holidays the place is swarming
with people who derive a lot of pleasure from some of the
animals' antics, who seemingly behave in front of them to
keep them amused. In fact, a librarian, who happened
to be chairman of an association to whom I was giving a
lecture, introduced me by saying, 'Our speaker is the
senior veterinary officer of London Zoo. In my library I
have several dictionaries, and I have checked up the
meaning of the word zoo.' He continued with a twinkle,
'A zoo is a large collection of wild animals under one
cover for the better observation of human behaviour.'
When you see how the animals behave to amuse the
public you can believe this, and also, when you see how
the public behave at times, you know full well that the
animals behave better.

The bird collection at the zoo always attracts a lot of

attention because there are plenty of beautifully coloured creatures who are very active, flying around, calling or shrieking. The parrot house is perhaps the best known of the bird collections; it's certainly the noisiest! The parrot family is a vast one and you can find examples in most tropical countries. Australia boasts several species, and the one that is probably best known is the sulphur-crested cockatoo. This is a fine big white bird with a lovely crest of yellow feathers on the top of its head, which it raises and lowers to express its emotions. It can also be trained to mimic the human voice and repeat phrases it hears repetitively.

One of the most popular exhibits in the parrot house was a sulphur-crested cockatoo called – wait for it – Cocky. Cocky had a habit of gathering a crowd around him by jumping up and down on his perch, screaming and shouting, and then moving to the edge of the perch and putting his head down to be scratched. Somebody in the public would oblige, at which point he would raise his head sharply, grasp their finger in his claw and give it a sharp nip. Nothing very serious, you understand, but enough to make the person yelp and the surrounding public roar with laughter. Then Cocky would also roar with laughter and jump up and down on his perch, wagging his sulphur-crested plume. Cocky was undoubtedly the noisiest bird in the collection.

Before the public arrived in the morning, the keepers would sweep and clean inside the aviary. He used to scream and shout and raise a fearsome din while they were trying to do their work, and they used to shout back at him, 'What do you want now?' Cocky would jump up and down and roar with laughter again. It was a game the keepers played and thoroughly enjoyed, and so did Cocky.

One day a member of the public who tickled Cocky's head wasn't quite quick enough to move their finger, and Cocky wouldn't let it go. The owner dragged his leg out through the bars of his cage, and, unhappily, fractured the parrot's femur in several places, causing shards of bone to pierce the skin. It was a bad compound fracture.

Poor old Cocky was about forty years of age at the time and was admitted to the zoo's hospital. He was given a light anaesthetic and, while he was sleeping peacefully, his leg was cleaned up and X-rayed, together with the good leg. When we developed the X-rays we were absolutely horrified at what we saw. If such a break had occurred in a human the leg would have had to be amputated.

We decided, however, to try and gum the leg together and use a pinning technique. This simply means pushing a rod up the marrow cavity in the fractured bone, joining all the pieces together so that they can heal.

There was plenty of literature on the sorts of devices

you could use to pin human bone, but there were no parallel pieces of equipment for birds. Birds have very fragile, hollow bones, so that they can fly. If their bones were as solid as ours they would never get off the ground. We had a delicate task ahead to find something suitable to fit inside the marrow cavity that would be firm enough to hold the fragments of bone together but not so heavy as to cause damage.

We thought and thought, and in the end we decided that the best form of pin would be a large hypodermic needle. We measured the width of the marrow cavity on the X-ray, and that of the appropriate hypodermic needles, then we cut the stainless-steel needle down to the right length. However, then we were baffled as to how to introduce the needle into the leg through the parrot's very complicated knee joint. Further consideration led to the idea that we should use a straight stainless-steel surgical needle, because a surgical needle has an eye in the top, like an ordinary sewing needle, through which we could thread a sufficient length of nylon cord to allow us to pull the rod up the bone into position and then withdraw the nylon thread, leaving the solid-steel prosthesis in position.

It worked like a charm. We were able to complete the operation successfully, and Cocky is the only parrot I know of to have healed and gone back on exhibition with a steel pin in his thigh bone.

When Cocky came round from the anaesthetic he was a little shocked because of the length of time the operation had taken, so we had a special recovery box made. It was warm, and a steady flow of oxygen-loaded air was pumped through it. Cocky was placed in the recovery box and he lay on the floor with his wings spread out and his head on one side, sleeping peacefully.

Quite suddenly, and unannounced, Prince Philip, the president of the society, arrived on a routine visit. It would, I feel sure, have pleased him to find us fully occupied and in action, having just completed an operation. I was able to show him round the hospital and demonstrate what we did and how we did it, and we finally finished up in the operating theatre where we had recently been operating on Cocky.

I asked the prince if he would like to see the X-rays of the bird, having explained to him what had happened.

'Yes please,' he said.

We went into the dark room and I displayed the X-rays of Cocky's leg, both before anything was done and afterwards, with the steel pin in position and the bone realigned. He asked question after question, and was only satisfied when he could repeat exactly the technique we had used. As we came out, I said, 'Would you like to see the patient?' He was very eager, and we took him over to the special 'hotbox' we had constructed. We all stood round, leaning over while I opened the glass lid. There, lying on the floor apparently peacefully asleep, lay Cocky. Prince Philip leaned over further and stared with the greatest of interest at the recumbent bird.

The apparently unconscious Cocky, lying as he was on his side, opened one yellow beady eye and stared back. Then, quite suddenly, he screamed out, 'What do you want now?' echoing the keepers.

The effect was dramatic and memorable: the prince threw back his head and roared with laughter, as did his aide, until the tears streaked down his face. He turned to his aide and said, 'I can't wait to tell my wife this story, she'll never believe me!'

*Cocky recovered and went back on exhibition,
behaving as raucously as before.*

TAIL PIECE

27

LAST DAY

MY LAST DAY AT THE ZOO WAS AN EXTREMELY emotional affair. I held a farewell party in the hospital for my fantastic dedicated staff, headed up by the hospital superintendent, Alec Wilson, and his second in command, Tony Fitzgerald. They were both wonderful men, and were supported by equally dedicated staff. After the party, I toured the gardens for the last time, alone, to say farewell to my friends the animals, many of whom had been treated for a variety of ailments. I went first to the Mappin terraces to see my bear patients, who had caused so much trouble when they'd escaped. I walked slowly down the main walk, pausing at the aquarium to see the huge carp that I had sutured after he was rescued from some bushes in the Welsh hills where he had been dumped after being landed. Then I went on to the reptile house to see the python on whom I'd performed a colostomy.

I crossed over to the monkey house and was greeted by a chorus of whoops from excited chimpanzees and studious gorillas. My whistling friend, Sukie, the South American monkey, came to the front of her cage and extended her delicate hand to grasp mine, as if she knew I was leaving.

I went to the lion house and bade farewell to the big cats. I had given two of the lionesses Caesareans and one a claw removal. Nepti the tigress stopped, purred and rubbed on the bars as I stroked her head – I had delivered two of her litters. Then I proceeded to the llama paddock: one of the females, whose broken foreleg had been repaired by a complex external bone-pinning operation, strode boldly up and down. I passed the birds of prey and Goldie the eagle, then went under the eastern tunnel to the middle gardens, where the new giraffe paddocks had replaced my old flat, and on to the north gardens to see the puma, who boasted a beautiful new coat after having surgery to remove a rodent ulcer. The emotional strain was awesome.

I was returning slowly to the hospital to bid my final farewells when I caught the glaring, inquisitive eye of Icarus. Many years earlier a small cardboard box had been dumped overnight on the steps of the zoo's main offices. I was sent for because, when the box was retrieved by Hanson, the housekeeper, a beady-eyed, featherless head had popped out of the folded cardboard flaps to reveal the inquisitive form of a sulphur-crested cockatoo. Further investigations showed there wasn't a single feather on his entire body or wings. The box contained no note or explanation. Nothing.

The naked bird came with me to the hospital, where his kindly nature and comic behaviour endeared him to all of us, but he made a particular friend of me. He was, of course, christened Icarus – an obvious name for a featherless one. Six months of therapy, oily dressings and an external cage where the whistling winds encouraged feather growth and, lo! a covering of downy feathers grew all over his body and flight feathers emerged on his wings. However, he never managed more than one yellow feather on his head, which was all that remained of his so-called sulphur crest. This single feather was raised and lowered like a battle flag.

Icarus's inability to fly did not deter him from trying. He would clamber up a table leg in the hospital dressing room, foot over foot, hook his beak over the table top, haul himself up onto it and view the whole length of the table as a runway. He would then thunder down the table top, flapping his inefficient wings, and hurl himself off the end, still flapping furiously. A learning curve followed to the ground, but still he was quite undaunted. To our delight, the little creature would raise his solitary yellow crest feather, walk over to the table leg and climb up again to repeat his performance. In more leisurely moments, he would get up onto my shoulder while I sat at my desk and nuzzle my neck.

As I prepared to depart from the hospital, I looked at my downy friend and bade him farewell, remembering fondly his many antics. I packed up my personal belongings from the hospital and left for my home in Swiss Cottage. Teaching at the Royal Veterinary College would seem a million miles away from all this, and I wondered whether my first day at the college would be as dramatic as my last day at the zoo?

I drove away along the roads leading home, and, halfway, I pulled in and made a decision. I turned the car round and went back to the zoo hospital, parked and went in through the swing doors. Walking about on the floor of the main hall was the object of my return: a sorrowful-looking Icarus. Without compunction, I found a cage, popped Icarus into it, put him on the back seat of my car and drove down to the Royal Veterinary College, handing him over, for the time being, to the tender care of the nursing staff in the Beaumont Animal Hospital. I then went home and unpacked my belongings.

When I took up my appointment at the Royal Veterinary College, my first job was to visit Icarus, who was delighted to see me and climbed onto my hand and up my arm to sit on my shoulder. He became a very firm friend of the nursing staff, who cared for him wonderfully, and it was always a debate as to which of us the bird gave most affection. Whenever I looked at Icarus and admired his sang-froid and his ability to take on any particular situation, I realized just how much he had done to help me in the transition from my momentous career at the zoo, which had taken me all over the world, to my new one as a teacher.

A Snowflake in My Hand

Samantha Mooney

A Snowflake in My Hand is Samantha Mooney's heartwarming account of the many years she spent working at New York City's famous Animal Centre.

In this unforgettable story, Samantha recalls the work of a dedicated team of professionals who, faced with tragic cases of abandonment and often incurable illness among the animals they cared for, somehow managed to find both companionship and laughter in that caring. As she looks back on how the centre became a sanctuary of hope where miracles sometimes happened, she also remembers one special, tiny black cat called Fledermaus who captured her heart and showed her the risks and the rewards of daring to love again.

'This book is not only for cat lovers, it is for lovers of life'
WASHINGTON POST